Human Awareness, Energy and Environmental Attitudes

Boris Aberšek • Andrej Flogie

Human Awareness, Energy and Environmental Attitudes

Springer

Boris Aberšek
Faculty of Natural Sciences, Research
Institute for Contemporary Technologies
University of Maribor
Maribor, Slovenia

Andrej Flogie
Faculty of Natural Sciences, Research Institute
for Contemporary Technologies
University of Maribor
Maribor, Slovenia

ISBN 978-3-031-05870-7 ISBN 978-3-031-05871-4 (eBook)
https://doi.org/10.1007/978-3-031-05871-4

This Springer imprint is published by the registered company Springer Nature Switzerland AG
The registered company address is: Gewerbestrasse 11, 6330 Cham, Switzerland

Abstracts

The basic structure of the book consists of 8 chapters.

Chapter 1, "Introduction," deals with issues such as the growing worldwide demand for energy, problems of scarcity, and environmental impacts associated with conventional energy resources.

Chapter 2, "A Philosophical View," discusses philosophical and societal relations associated with energy and environmental issues.

Chapter 3, "Environmental Costs Hierarchy," is concerned with the human impact on the environment.

Chapter 4, "Social Relations," addresses the development of social competences and critical thinking.

Chapter 5, as the title suggests, is devoted to energies and energetics.

Chapter 6, "Green Solutions," discusses energy resources that don't harm the environment through factors such as releasing greenhouse gases into the atmosphere.

Chapter 7, "A View to the Future," focuses on already-existing, alternative, and under-researched or unaffordable solutions.

Chapter 8, "Conclusion," sums up the main points presented throughout the book.

The references used during the creation of the book are listed under "Literature" after the final chapter.

Book Review by Matej Vesenjak

As the starting point of the problem, the authors point out that there are only three big monopolies in the world: *food*, *energy*, and *technologies*. Today, in one way or another, all three are narrowly connected to technologies. Historically, humans have used technology to give acquire a certain way of life. Generally, humans don't want to change this comfortable lifestyle. However, to maintain the way we live, energy is *needed,* and its production has a drastic impact on our environment. This scientific book indicates the original sin of the current environmental problem and sustainable development and the necessity to change the thinking, that is, the lack of criticality and critical judgment of society as a whole and of individuals as its fundamental elements. The authors point out that this uncritical way of thinking also results in various misleading information and delusions, which may be unintentional, arising from ignorance and misunderstanding of the problem, or intentional, resulting from a desire to manipulate society.

This book, *Environmental Awareness and Energy and Education*, presents four basic areas, divided into nine chapters. Thus, the authors indicate in the introduction the current problems of energy production and our attitude to the consequences of these problems on environmental sustainability. It underlines the need for environmentally conscious engineering and stresses that environmentally conscious engineering thus incorporates important technical and analytical techniques; best practices for the design of industrial, commercial, and consumer products; and processes that are environmentally friendly and comply with environmental regulations. The authors point out that this can be achieved by incorporating a systemic approach into the design process and

1. Different AI intelligence-based optimization models
2. Different design practices for the environment, guidelines, methods, and tools
3. Assessment and design for the overall life cycle of a single product based on a systemic approach
4. Reverse engineering

5. Environmental management systems
6. Taking into account the various standards, as ISO 14000, as well as just emerging standards or standards that are yet to be created in the future, and much else.

In the second part, the authors discuss social relations related to energy, environmental issues, sustainable development, and environmental awareness. The second part is divided into four chapters, in which the authors focus on the philosophical, sociological, and psychological views of the question and analyst how the individual and society perceive, process, and analyze the information obtained. The book is rounded off with Conclusion, where the consequences of inaction and possible solutions are analyzed, as well as Applied Literature and Index.

This scientific book highlights extremely topical problems and suggests solutions to these problems, focusing mainly on the long-term causes and consequences of environmental problems. On the basis of what has been written, I believe that the content presented in the scientific book represents an essential contribution in the interdisciplinary field of environmental awareness and sustainable development, thus expressing my positive opinion.

Faculty of Mechanical Engineering,
University of Maribor, Maribor,
Slovenia

Matej Vesenjak

Book Review by Stanislav Avsec

The book begins with the thesis that in today's technologically advanced, rapidly evolving world, we need to shape our thinking more and more quickly (today's information is already trashed tomorrow) based on:

1. Lack of information
2. Information requiring a huge amount of narrow and specific expertise, which is completely incomprehensible to most and often inaccessible to most
3. Misleading information (unintentional and intentional)

Let's focus only on the latter, which is misleading information. As we mentioned, these can be unintentional and deliberate.

1. *Inadvertently* we call that misleading information that arises when certain 'experts' (politicians, influencers, etc.) have to create (and then communicate) their opinion in the absence of relevant knowledge. That's how they have to make decisions under System 1 (Kanemann), in a hurry. However, since they also lack the experience on the basis of which they would make at least a somewhat competent decision, the likelihood of making the wrong decisions (according to Murphy) is extremely high, inappropriately greater than this, their decisions and then the information about them will be correct.
2. However, misleading information can also *be intentional.* They are passed by people who have certain interests and want to manipulate them deliberately. These people have enough information, correct information, but they distort the truth for reasons known to them. Deliberately misleading information is the hardest to detect and is the most dangerous – for every individual and therefore for society as a whole, because in order to achieve their partial interests, those who use it are willing to destroy everything around them.

The third part of the book talks about technological relations and is presented in four chapters. Here, the authors looked at different forms of energy production. Special emphasis is placed on the future. The chapter "Look to the Future" focuses on solutions that are now more or less well known, alternative, but not yet sufficiently researched or affordable. These are, for example, nuclear fusion as the most promising source of energy of the future, artificial photosynthesis, and other different modern visions of energy production. It also touches on the storage of energy and greenhouse (and other environmentally unfriendly) gases, the recycling and re-use of energy waste and their recovery, and the storage or reuse of various wastes from different sources and technological processes and, consequently, their effects on the environment.

The book is rounded by authors with Conclusion, Applied Literature, and Index.

The book highlights extremely topical problems and suggests solutions to these problems, focusing mainly on the long-term causes and consequences of environmental problems.

Faculty of Education, University of Ljubljana, Ljubljana, Slovenia

Stanislav Avsec

Preface

Young brains come to school to learn and developed awareness; older brains come to school to teach them. At least, that's the ideal. The rationales for writing this book were the questions: *can we tell anything about how young brains learn and developed awareness from older brains, and therefore how older brains should teach younger brains for optimal effect*? and, additionally, *what is most important in this process?*

The system of knowledge between younger and older brains changes in such a way as to distribute the knowledge between the two. By way of analogy, a similar thing can be said about the system of transferring the knowledge between one brain (the teacher) and several brains (students in a class) in some way of communication, which is what goes on in traditional school settings. However, if we focus only on environmental issues, the existing relationship between the young brain and the old one shows a fundamental mistrust, which is clearly seen in the statement of the young brains, namely the comments of young people across the world to attitudes of the old brain, or especially to the attitudes of the decision makers towards the environment have clearly shown what they think: "BLAH, BLAH, BLAH." On behalf of these youngsters, Greta Thunberg criticized leaders over the climate crisis, saying:

> *"They invite cherry-picked young people to meetings like this to pretend that they listen to us. But they clearly don't listen to us. Our emissions are still rising. The science doesn't lie".*
>
> *"We can no longer let the people in power decide what is politically possible. We can no longer let the people in power decide what hope is. Hope is not passive. Hope is not blah, blah, blah. Hope is telling the truth. Hope is taking action. And hope always comes from the people."*[1]
>
> *And the fundamental question arises: How could (must) change this mistrust?*

Maribor, Slovenia

Boris Aberšek
Andrej Flogie

[1] https://www.theguardian.com/environment/2021/sep/28/blah-greta-thunberg-leaders-climate-crisis-co2-emissions.

Acknowledgments

The author gratefully wishes to acknowledge:

The Ministry of Education of the Republic of Slovenia, ESF, Slovenian Research Agency, and many Erasmus+ projects for ideas, research, and their financial support.

All of our colleagues from the Faculty of Arts, the Faculty of Natural Sciences and Mathematics, and the Faculty of Education, University of Maribor, Slovenia, and from the Institute Anton Martin Slomšek, Maribor, Slovenia, for their contribution and helpful comments. Thanks to all our colleagues for all our conversations, in which we invented questions and jointly examined our intuitive answers. We are greatly indebted also to all our students, for their inspiration and assistance. Many thanks to Helena Fošnjar for translation and proofreading.

The author wishes also to express his appreciation to all those who have ensured the quality of this book. Last but not least, thanks to our parents, partners, and children, who have inspired us, supported us, and given us the opportunity to be who we are.

Contents

Chapter 1
Introduction

Abstract The first chapter points out current issues associated with energy extraction and explores human attitudes toward the consequences this has on the environment. Environmental problems directly related to energy production and consumption include air pollution, climate change, water pollution, thermal pollution, and solid waste disposal. The chapter begins with three famous quotes and draws on the ideas expressed by these historical personalities:

Lupus est homo homini, non homo, quom qualis sit non novit (Plautus).

Homo, sacra res homini (Seneca).

Before the Law: 'No,' said the priest, 'you don't need to accept everything as true; you only have to accept it as necessary.' 'Depressing view,' said K. 'The lie made into the rule of the world' (Kafka, The Trial).

A man doesn't need an outside enemy; he's the greatest threat to himself. But he's not just a threat to himself; he's a threat to everything around him as well. No living thing on Earth destroys it more than it takes for its existence, except for human. And for that, he's willing to do anything, deceive, and kill, to destroy and to justify what it does with a backlash and a lie.

Lupus est homo homini, non homo, quom qualis sit non novit (Plautus).

Homo, sacra res homini (Seneca).

In front of the law: "No," said the priest, "you don't need to accept everything as true; you only have to accept it as necessary." "Depressing view," said K. "The lie made into the rule of the world" (Kafka: The Trial).

A man doesn't need an outside enemy; he's the greatest threat to himself. But he's not just a threat to himself; he's a threat also to everything around him. No living thing on Earth destroys it more than it takes for its existence, except for human. And for that, he's willing to do anything, deceive, and kill, to destroy and to justify what it does with a backlash and a lie. But why? Let us start our stories with some basic truths.

B. Aberšek, A. Flogie, *Human Awareness, Energy and Environmental Attitudes*,
https://doi.org/10.1007/978-3-031-05871-4_1

There are only three major monopolies in the world:

- Food.
- Energy.
- Technologies.

Today, all three monopolies are closely related to technologies in one way or another. Technologies penetrate our everyday life; we cannot even begin to imagine modern agriculture without technology, and, needless to say, the production of energy today is also based on many different kinds of technologies. Since its very beginnings, mankind has tried to make life easier by way of technology, to control and change nature by means of technology – of course, according to man's own desires and visions. Technology has made man's life easier; it became a tool for managing and mastering nature. At first, man was technology's humble assistant; later he became its master, only to now finally become its slave: we can no longer imagine our everyday life without all these different kinds of technologies that surround us in this day and age. Moreover, technology has not only enslaved man; rather, it has already reached a developmental stage of autonomy in which the role of man is becoming increasingly irrelevant and probably even unnecessary. Nevertheless, let us focus on just one of the three major monopolies today, the one that is crucial for the existence and growth of the other two – energy. Without energy, there is no technology, nor food production, as we know it today.

In the course of human development, technological progress has enabled a certain way of life for people. No one would want this to change; no one would like to see our way of life taking a turn for the worse. *However, in order to maintain our modern lifestyle, we need energy.* Fundamentally important in this respect are energy conversion technologies and the sources from which we obtain this energy. The conversion cycle is a process ranging from:

- Carriers or sources of energy, including energy products in which different types of energy are stored.
- Conversion energies.
- Useful forms of energy. Of course, useful forms of energy are defined anthropocentrically; they refer to forms of energy that are beneficial to humans. These are the following:
  - Light.
  - Heat.
  - Mechanical work.
  - Chemical energy.

A central question in all of these processes is in what way we are able to carry out these transformations, i.e., we are talking about the efficiency of the processes of production and extraction of useful forms of energy. Unfortunately, we are still at a very primitive level in this respect: we are able to convert energy sources into useful energies only with a very low efficiency (up to approximately 10%), whereby the lower the efficiency, the greater the consequent "inefficacious" environmental impacts.

To arrive from primary forms of energy (energy sources) to forms of energy that are useful to humans, we use a variety of technologies, mainly depending on the individual energy source (or product) in question. There are two elements, which are essential to these conversions:

- The first is *economic:* What kind of energy source (or product) are we using? How many losses occur in the process of conversion? In other words, we are concerned with efficiency in such processes. This, of course, is closely linked to the cost of producing energy and the price of that energy, which we as individuals have to pay.
- The second aspect is *environmental:* How do the technological processes of energy conversion affect our environment? This is a universal societal issue, a problem that we do not perceive directly in the sense of costs (expenses paid for energy products) but only indirectly through the responses of our environment to our immediate life. Two options are possible here:
  - Either we acknowledge the changes affecting the environment (admitting that both individually and collectively we are contributing to this problem).
  - Or we deny these changes, hiding behind more or less logical explanations as to why these changes are occurring. The common denominator of such justifications (excuses) is that these changes are not our fault but someone else's.

In this book, we will not focus on the economic aspect but rather on the second one, more specifically, on the impact of the production and use of energy on our environment, as well as on the human attitude toward these influences. We will pursue two central goals:

1. *To raise awareness* in the people who deny these changes, explain to them the issue in such a way so that more and more people will understand it and, in turn, become part of the solution.
2. To move toward *shifting* the entire society from the short-term goals of individuals and consequently cheap energy to the long-term goal of preserving our environment for the next generations. In pursuing these goals, we need to consider this issue in a humane, systemic, comprehensive, and, above all, long-term manner, i.e., in a sustainable way.

In a systemic approach, an individual needs to be considered as a building block of society, since each individual can also be part of a decision-making system, an employer, a shareholder, a politician, or anyone who affects the society at a general level.

Two general "solutions," i.e., potential strategies, are used as a starting point in this book:

1. Use of environmentally friendly energy sources, i.e., green technologies.
2. Reducing energy consumption and consequently reducing the pressure on the environment, which can be achieved by the following:

- Saving energy and changing our way of life.
- Using technologies and energy sources that are less harmful to the environment.
- Improving the efficiency of existing technologies (which is the current trend).
- Developing new technologies based on existing energy sources but with a better efficiency (e.g., fourth-generation nuclear fission reactors).
- Developing completely new ways of energy extraction (e.g., nuclear fusion). This is known as environmentally conscious engineering, which focuses not only on an environmentally friendly production of energy but also on a proper understanding and implementation of these processes.

Environmentally conscious engineering includes important technical and analytical techniques and best practices in the design of industrial, commercial, and consumer products and processes that are environmentally friendly and comply with environmental regulations. Environmentally conscious engineering can be achieved by including (in this systematic process) the following:

- Various optimization models based on artificial intelligence.
- Various practices, guidelines, methods, and tools for environment design.
- Relevant assessment and concepts for a comprehensive life cycle based on a systematic approach.
- Reverse engineering.
- Environmental management systems.
- Considering various existing standards, e.g., ISO 14000, as well as emerging standards or standards that will occur in the future and much more.

Overall, in the short-term perspective, as individuals, we have the possibility to change our way of life and reduce energy consumption, or we can make a decision as a society to use more expensive energy sources and pay more for our way of life. Either way, making the decision is not easy, because, as construal level theory[1] (CLT) (Trope & Liberman, 2010) suggests, the more distant an object is from the individual, the more abstractly it will be thought of, while the closer the object is, the more concretely it will be thought of.

The individual decision depends on each person, while the collective decision is related to the idea of social consensus, to the individual as a constituent part of the society as a whole. Both decisions are difficult: on the one hand, *we have clear and well-understood short-term goals* (to live better now), and on the other, we have wider social goals, which are more long-term and perhaps more "blurred" (e.g., to make it possible for the generation of tomorrow to benefit from the same things that were available to the generation of today). In order to achieve either or both of these goals, our individual and collective social awareness both need to change: *we need to* revise our value systems; develop social competencies, emotional intelligence, and affective relationships; and reevaluate our social psychology. Such changes would

[1]CLT describes the relation between psychological distance and the extent to which people's thinking is abstract or concrete.

affect an individual's *psychological distance*, which assumes that the more distant in time (and in space) the individual's objective is, the more abstractly he will perceive it. If the objective is closer, it is perceived more concretely, it is more easily understood, and an individual identifies with it more clearly as well. There is also another option – environmentally conscious engineering. This is a slightly more long-term and utopian alternative, which is associated with discovering new energy extraction technologies (e.g., artificial photosynthesis) or developing already existing technologies (e.g., nuclear fusion) to a commercially viable degree.

Long-term and more or less utopian solutions will be indicated in this book; however, we will focus mainly on medium-term goals (goals which we see as achievable by 2050), emphasizing in particular the things which we can and ought to do today for the day after tomorrow! Why now? A brief look at the statistics of our planet's environmental footprint, which reflects the rate of resource use and the amount of waste generated and compares it to how quickly nature can absorb waste, create new resources, and remediate the consequences, reveals that we are spiraling inevitably toward doom. This is reflected in all segments of society, from problems in waste disposal and cleaning to the unstoppable pollution of air, water, and land. Hence, the time to address these issues is now.

This book is divided into nine chapters. The *Introduction* points out current issues associated with energy extraction and explores human attitudes toward the consequences this has on the environment. The second chapter, *A Philosophical View*, discusses philosophical and societal relations associated with energy and environmental issues. This chapter is divided into separate sections focusing on the philosophical, sociological, and psychological views and providing an analysis of how the individual and the society perceive, process, and analyze information on this matter. We point out that, based on CLT and from the point of view of psychological distance, we need to:

1. *Raise awareness* and bring this topic closer to the reader, so that he/she will reflect on these issues and perceive them less abstractly and more concretely.
2. *Raise social competence and emotional intelligence* so that one is able to place long-term societal goals before one's own short-term goals, not thinking only about today but also creating a vision of the future.

The third chapter, entitled *Environmental Costs Hierarchy*, deals with the human impact on the environment, especially in terms of the ecological footprint, and the related concept of human ecological awareness.

Chapter 4, *Social Relations*, addresses the development of social competences and critical thinking and points out that in today's technologically developed world, we need to develop our (critical) thinking more and more quickly (today's information is worthless tomorrow). This is related to two basic paradoxes, namely:

1. The problems addressed are increasingly more complex and more interdisciplinary, while our actions are increasingly more monodisciplinary.
2. Providing solutions to such problems requires time, but in today's quickly developing world, time is what we don't have.

However, on every step, we are confronted with imperfect information, information that requires a huge amount of specific expertise, or misleading information (whether unintentionally and intentionally). In the world of false prophets generating misleading information, there are countless believers (in the language of the Internet, followers) who blindly believe everything these false prophets have to say, as indeed, *true believers* would.

Using a social-cognitive perspective, researchers can study a wide range of topics including attitudes, person perception and person management, social awareness and social management as a part of emotional intelligence, prejudice, stereotypes, self-concept, discrimination, persuasion, decision-making, and other areas.

Chapter 5, *Energies and Energetics*, and Chap. 6, *Green Solutions*, discuss various forms of energy extraction and production. In the following section, special emphasis is placed on a *view to the future*. This chapter focuses on already existing, alternative, under-researched, or unaffordable solutions, such as nuclear fusion (the most promising future energy source), artificial photosynthesis, or various other modern visions of energy extraction. We also touch upon topics such as the storage of energy and greenhouse and other gases, recycling and reuse of energy waste and its processing, the storage or reuse of wastes from different sources, or technological processes and their impact on the environment. In this chapter, we summarize the theoretical ideas and knowledge from previous chapters, and we no longer deal so much with the concrete ecological problems themselves as with trying to suggest potential solutions, based on critical thinking and rational judgment. In doing so, we focus in more detail on possible solutions for two of the most pressing problems in terms of sustainability and the environment:

1. The problem of using "alternative," nontraditional forms of energy production and resources. We will highlight the critical relationship toward the use of nuclear energy among its opponents and advocates and attempt to demonstrate how we can develop critical thinking and a critical attitude toward certain issues at a systemic and long-term level, with emphasis on environmental issues and the ecological footprint.
2. The problem of transport, the related stress on the environment, and the impact of transport on the ecological footprint. We will highlight the modern trends for a transition to using electricity in transport. The process is reverse to the one described above: from a global problem of "even" pollution (e.g., exhaust gases of internal combustion engines), we are trying to concentrate locally all emissions to electricity production. We should not forget that electricity is still mostly obtained from fossil fuels, if we disregard hydropower plants and nuclear reactors (at least until the year 2050/2060).

In both cases, we focus mainly on the causal links between energy production and energy consumption and the impacts of these processes on the environment and the ecological footprint. These issues are observed mainly from a long-term perspective, from the point of view of sustainability, environmental awareness, and a possible change in the attitude of the individual and the society as a whole.

We round up the book with the sections *Conclusion*, *References*, and *Index*.

Chapter 2
A Philosophical View

Abstract The second chapter, *A Philosophical View*, discusses philosophical and societal relations associated with energy and environmental issues. This chapter is divided into separate sections focusing on the philosophical, sociological, and psychological views and providing an analysis of how the individual and the society perceive, process, and analyze information on this matter. Based on the construal level theory and from the point of view of psychological distance, we need to *raise awareness* and bring these topics closer to the reader, so that they will reflect on these issues and perceive them less abstractly and more concretely. Further, we need to *improve social competence and emotional intelligence* so that we will be able to place long-term societal goals before one's own short-term goals, thereby not thinking only about today but also creating a vision of the future. The global environment, planet Earth, is a highly complex system, which we cannot truly understand with the help of simplified models that we, humans, are able to process. Our human understanding of the complexity of a grain of sand simply isn't enough to construe the universe. This would require time and knowledge, which we do not have today and which we will probably not have for a long time.

> The environment is a system, a complex system, which we are unable to comprehend, let alone truly understand, with the help of simplified models that we, humans, are able to process. Our human understanding of the complexity of a grain of sand simply isn't enough to construe and adapt the universe! This would require time and knowledge, which we do not have today and which we will not have for a long time.

How do people think and perceive truth(s)? How do we respond to these truths? What kind of knowledge is necessary for understand the problems? What kind of knowledge is considered welcomed and desirable, even if it is not really necessary, and when does such knowledge stop being a means of raising awareness (and thereby addressing a problem)? We are talking about the human affective realm, or that part of Bloom's taxonomy, which is referred to as the affective domain

B. Aberšek, A. Flogie, *Human Awareness, Energy and Environmental Attitudes*,
https://doi.org/10.1007/978-3-031-05871-4_2

(Bloom et al., 1956). This concept is closely related to the notion of emotional intelligence, which was defined much later by Joel Robert Davitz, Michael Beldoch, and most noticeably by Daniel Goleman in his work *Emotional Intelligence: Why It Can Matter More Than IQ* (Goleman, 1995). It is important to highlight the individual as such and their affective values, which are formed mainly inside their immediate environment, and, complementarily, the individual as part of society, their social values, which are a reflection of the entire society's influence on that person's value system. In this book, we will concentrate primarily on issues related to environmental awareness and the human-society-environment relation. Let us begin by considering the ways in which our behaviors are shaped, by looking at *what we want to believe* and *what we can believe* (Goleman, 2007).

1 What Is True and What Isn't?

In a world increasingly beset by fake news, it is becoming ever more challenging to establish what is true and what isn't. At first, let us try to "scientifically" define and substantiate what *reality* is. In attempting to do so, we must distinguish between *reality* and the *real*. As Lacan explains, reality is an external reality, our social and tangible space, which we are accustomed to, we are able to find our way in it, and interact with others. By contrast, *the real* is invisible, a phantom entity, and therefore seemingly omnipotent. At the moment in which this phantom agent becomes part of our reality (even if this refers to, for example, changes in the weather, such as storms, floods, and droughts), its power is localized, and it becomes something we are able to confront (even if we are defeated in this confrontation). Until this reality is transmitted, interpreted, and internalized, "we are caught up in an anxious paranoia (pure globality) or resort to ineffective symbolization with stagings, exposing ourselves to unnecessary risk (pure locality)." These "ineffective symbolizations" have already taken on various forms. Among the most well-known is probably Donald Trump's disregard for climate changes and the risk of not taking any action. A similar thing happened with the emergence of the coronavirus pandemic in 2020. Such actions are far worse than shouting and cheering while watching a football match on television at home, acting as if we could miraculously affect the outcome of the game. But this does not mean we are powerless: we can get out of this grip even before science offers the technical means to hold back climate changes. In Lacan's terms, you should try to identify yourself with your symptom and do so without any embarrassment – or, in simpler terms, take full advantage of all the small rituals, forms, whims, and anything that supports you in your daily life. Everything is allowed, which could work to prevent a mental breakdown, including various forms of fetishistic denial: "I am well aware… (of how serious the situation is), but still… (I don't believe it completely)." Don't think too much in the long run – just think about today. If it helps, play a game from *Life is Beautiful:* pretend that climate change is just a game you play with your family voluntarily, and if you win, you'll get a huge reward. And, while we're speaking of movies (if you have any spare time for watching films), succumb freely to all your temptations: catastrophic anti-utopias, daily life with TV sitcoms, and a podcast on YouTube about the world's deadliest natural disasters.

2 Science and Reasoning

Let us make a reference to the scientific interpretation of physical quantities, which suggest a possible explanation of what is (or what might be) *real*. Before we embark on a discussion about reality and a scientific approach to its definition, we need to define some basic concepts from science and answer some fundamental questions about what science is:

What is science? This question can be viewed from two perspectives, namely:

- *A commonsense perspective*
  Science stands for subjects (subject areas) such as physics, chemistry, biology, mathematics, and technology (i.e., STEM subjects) – but not the arts, music, or theology, for example.
- *A philosophical (scientific) perspective*
  This view focuses on the characteristics and methods which all the above subjects (subject areas) have in common – i.e., *what is it that makes something science*.

If someone wants to think scientifically, like a scientist (an expert in a certain field), then science needs to be considered from a philosophical perspective, which immediately raises the question:

- Is science not merely an attempt to understand, explain, and anticipate the world in which we live? Perhaps. But, is that all science is?
- The question of what science is, therefore – much like many philosophical questions – is more complex than it appears at first sight.

A quick answer to these questions might be as follows:

- Science is distinguished by its *specific methods*, which scientists apply to explore the world.
- *Science forms theories*. One of the key issues in the philosophy of science is understanding how techniques such as experimentation, observation, and theory formation have allowed scientists to uncover so many nature's mysteries.

Science, then, is marked by a typical way of thinking/reasoning – *the scientific thought*.

2.1 Scientific Thought

Scientists (experts) often communicate to us things about the world, which we would otherwise never think about, consider, or believe, for example:

- Biologists tell us that we are related to chimpanzees.
- Geologists tell us that Africa and South America were once part of a single landmass.
- Physicists and cosmologists tell us that the universe is expanding.

But how do scientists arrive to these very amazing realizations? The answer, again, appears simple at first sight:

By thinking and reasoning!

But what is the nature of scientific thought? To what extent can we trust the conclusions of scientists, or, even harder, to what extent can we trust the conclusions of pseudoscientists, or influencers, in today's world that is riddled with fake news? The easiest way is to start thinking like actual scientists/experts and to start using the same or similar methods and ways of reasoning. There are a number of methods out there; let us consider here only the most often used ones:

- *Deductive reasoning*, from the more general to the more specific. If something applies to everyone, then it is going to apply to me as well.
- *Inductive reasoning*, from specific observations to broader generalizations. If something applies in specific cases, then it may (must) also apply in general, e.g., the sun rose every day so far, which means it is going to rise tomorrow as well.
- *Inference to the best explanation.* (Conclusions are reasonable and offer a better explanation than other *interpretations.* This is the best available explanation of facts on the base of probability.)

2.2 Argumentation

Once we have arrived, in some way or another, to certain conclusions, we need to confirm (or reject) these conclusions; we need to provide argumentation. Argumentation tries to take effect on an audience, to change the opinion or atmosphere of the audience and reach a consensus (a compromise through conversation), and not to enforce an own will by means of coercion or dressage. In science, this is where we use the old maxim: "use the power of argument rather than arguments of power."

Our findings can be confirmed by formally correct argumentation:

- Empirical argumentation, through empirical research.
- Statistical argumentation based on actual data processing, through performed methodological research.

- By referring to valid covering laws and theoretical findings and through argumentation (persuasion).

  Three kinds of arguments are known in this respect:

- Arguments of the type *ethos* (these are "arguments of power"). For example, "I am your professor, I know how things stand.../I'm an expert on this matter..." (pseudoscientists, influencers), etc.
- Arguments of the type *pathos* (playing on emotions). For example, "unless we act right now, glacier erosion will continue irresistibly."
- Arguments of the type *logos*. These are logical arguments based mainly on the above described (logical) ways of scientific thought.

The difference between *common argumentation* and *formally correct argumentation* is that in formal correct argumentation we use signs, which are supposed to be entirely unambiguous, while common argumentation occurs in the natural language, where ambiguity is not excluded a priori and, in fact, it is often deliberately included.

For persuasion (speech, writing, performance) to be truly effective, three factors need to be taken into account, which Aristotle describes succinctly using three terms:

Ethos (personality, charisma, reputation). Appeal to the audience through affection.
Pathos (passion, emotions). Each word, not heard by the soul, is lost.
Logos (logic, language, the mind). Commonsense reasoning.

In most cases, scientific thought and argumentation are based on common sense. Hence, let us focus only on the latter out of the three, the *logos*, or logical (scientific) argumentation/reasoning.

2.3 Scientific Explanation

What exactly does it mean to say that a certain phenomenon can be "explained" using science?

2.3.1 Carl Hempel: Covering-Law Explanation Model

Scientific explanations are usually offered as satisfactory answers to what Hempel called "explanation-seeking why questions." If we could determine the essential characteristics, which such an answer must contain, we would be able to understand what a scientific explanation is.

Hempel suggests that scientific explanations typically have the logical structure of an argument, a set of premises followed by a conclusion. The conclusion states that the phenomenon that needs explaining (in our case, climate change, global warming, or environmental footprint) actually occurs, and the premises tell us why the conclusion is true.

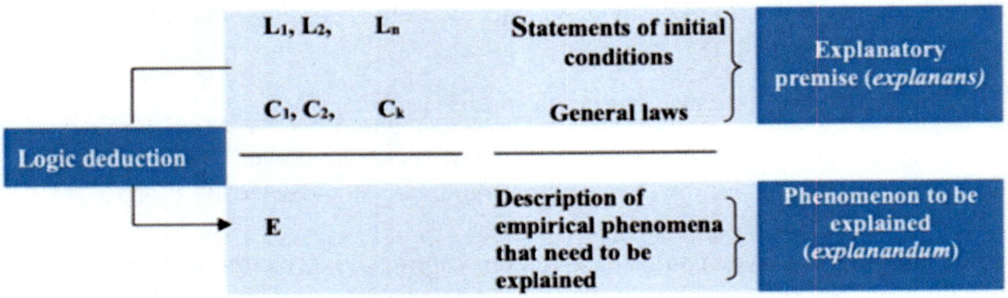

Fig. 2.1 The form of Hempel's and Oppenheim's explanation of the covering law

Hempel's covering-law model addresses an explanation as a deduction from laws in conjunction with views on preconditions. Hempel defines this as follows:

> *[The covering-law] explanation answers the question: Why do explanatory phenomena occur at all?; by showing that the phenomena arise from certain particular circumstances $C_1, C_2, \ldots C_k$, depending on certain laws $L_1, L_2, \ldots L_r$. By exposing this, the argument shows that, given the particular circumstances and the laws in question, the occurrence of the phenomenon could be expected; and it is in this sense that the explanation enables us to understand why the phenomenon occurred* (Hempel, 1965, 337).

Based on the well-known position expressed by Hempel and Oppenheim (1948), explanation has the structure of valid deductive arguments, in which the explanation is a connection of *two premises* – a *statement of initial conditions* and a *statement of general laws* – and the explanation is a conclusion describing the empirical phenomenon, which we wanted to explain, which is shown in Fig. 2.1.

2.3.2 Example: Reality-Space-Time

Space and time are one of the most exciting areas of human exploration, as they form the arena of reality, the substance from which the universe is made. Everything we do, think, and experience happens in a specific part of the space and at a specific time (e.g., melting glaciers today). And yet, science still does not fully understand what space and time represent and how they could be explained with the help of science. Are space and time real physical quantities (of the type "logos"), or are they just useful ideas; are they part of the material world or the world of ideas? And if they are real, material quantities, are they basic, independent quantities, or do they derive from even more fundamental components? What does it mean that the space is mostly empty? Does time have a beginning? And last but not least, can we really control space and time? In what follows, we will try to outline very briefly the 300-year-old efforts to provide answers to these basic questions. We will touch upon the definition of *state-space* of dynamical systems, in which time plays a key role (Green, 2005). Figure 2.2 shows the connections between the disciplines that create the appearance of reality in the contemporary science of today's world.

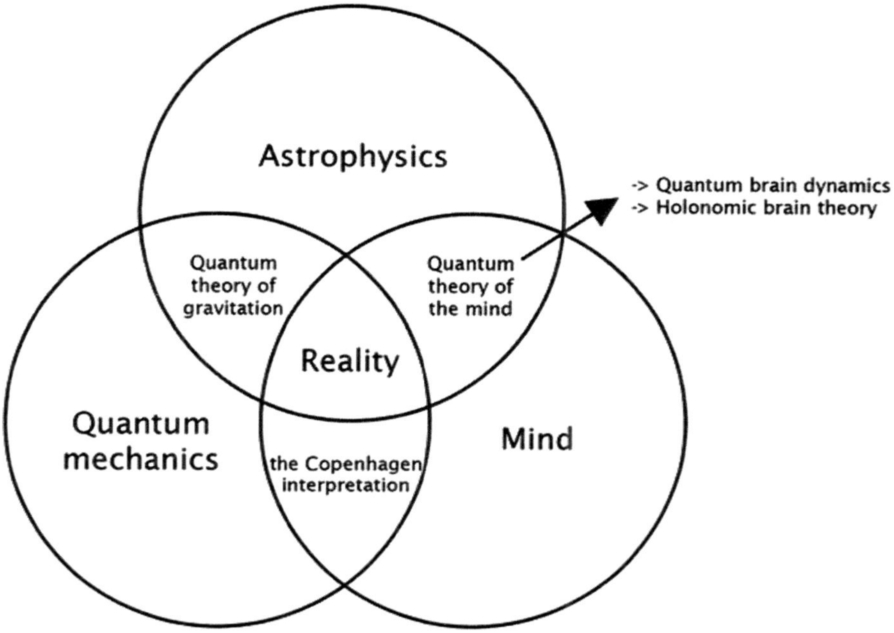

Fig. 2.2 Connections between disciplines that create reality

In any type of scientific research, one is always faced with a broad but very elusive question: what exactly is reality? How do we know that any assumption, idea, theory, or mathematical model of the system under study is a reflection of the true state of this system in real space and time? How can we, humans, who have access only to our inner perceptions and thoughts, be certain that these really are a reflection of the outside world? Philosophers have long since recognized this problem, while other scientists are aware that the world in front of us, the world we observe as it evolves in the arena of space and time, may not really have much in common with the desired reality. Especially over the last hundred years, scientific discoveries have led to findings that are as dramatic, awe-inspiring, and surprising, as only the most ingenious science fiction writing can be. But let us take things one step at a time.

Many of the questions posed above, perhaps in various guises, have already been asked by Aristotle, Galileo, Newton, Einstein, and many others, and so we are able to trace back the answers offered by one generation, refuted by the next, and improved and upgraded by modern science in the following centuries. Thus, for example, we can trace back the answers to the question of whether completely empty space is a real entity or a purely abstract idea. We can go back to Isaac Newton, who claimed in the seventeenth century that space is real, through Ernst Mach, who concluded in the nineteenth century that it is not, and to Einstein, who in the twentieth century dramatically reformulated the basic question itself, merging space and time and largely refuting the claims of his predecessor. Further on, by redefining the term "empty," the basic question of time and space arises anew yet again: apparently, the space is imbued with the so-called quantum fields and perhaps also with uniformly scattered energy, known as the cosmological constant – a modern echo of Descartes' old and refuted notion of an all-pervading ether. Our own era, finally, assumes that the space we inhabit is perhaps just a fraction of the much greater and more glorious cosmos.

As Algerian-born philosopher Albert Camus suggested, progress in physics, such as understanding the number of space dimensions; or advances in neuropsychology, such as understanding the organizational structures of the brain; or progress in any other number of scientific undertakings may fill in important details, but their impact on our evaluation of life and reality would remain negligible. Surely, however, reality is not what we think it is; reality is revealed to us only through our experiences (Camus, 1999). Through the scientific endeavors of recent decades, we have learned that human experience, our beliefs, is often a misleading guide to the true nature of reality, as just beneath the surface of the everyday there is another world we would hardly recognize, or at least we interpret it in each in our own way.

Over the last three centuries, advancements in physics have revealed space and time as the most baffling and most compelling concepts and also as the most instrumental concepts in our analysis of the universe and life in it. Such advancements have also shown that space and time are at the top of the list of age-old scientific constructs that are being radically revised by cutting-edge research. To Isaac Newton, space and time simply were that they formed an inert, universal cosmic arena in which the events of the universe played themselves out. To his contemporary Gottfried Wilhelm Leibniz, "space" and "time" were merely the vocabulary of relations between where objects were and when events took place, nothing more. To Albert Einstein, space and time were the raw material that underlies reality. Einstein's relativity theories shocked the way we think about space and time and revealed the principal part they play in the evolution of the cosmos. Ever since, space and time have become the crowning glory of physics, familiar and mystifying at the same time, and fully understanding space and time has become physics' greatest challenge (Green, 2005).

We are only going to discuss the basic laws of physics briefly, but our scientific endeavor will focus more closely on the philosophical foundations. Since the story of climate change is far from being fully written, we will not arrive at any final conclusions. We are going to encounter a series of developments – some intensely strange, some deeply satisfying, some verified by experiments, and some entirely speculative – that will show how close we have come to capturing all the possible scientific consequences and tackling the true texture of reality. Let us focus on the human-society-environment relation and take as our starting point *the relationship between the planet's biological capacity and the ecological footprint.*

3 How Do We Know, How Do We Believe?

Our knowledge and our perceptions are something that we experience through our perceptual world. However, one may ask oneself whether we can truly believe ourselves and truly rely on our perceptions, our own mental images. In order to prove this, philosophers use various tools in such situations, such as thought experiments. To illustrate this, let us consider John Searle's *Chinese room experiment* (Searle, 1980). Searle concocted it to convince us that computers don't really "think" as we do; they manipulate symbols mindlessly, without understanding what they are doing. Searle meant to make a point about the limits of machine cognition. Recently, however, the Chinese room experiment has motivated some researchers into reflecting on the limits of human cognition. Human beings can be mindless too, even when engaged in a pursuit as eminent as quantum physics or, in our case, the environmental footprint.

Searle first proposed the Chinese room experiment in 1980 (Searle, 1980). At the time, some artificial intelligence researchers claimed that machines would soon pass the Turing test,[1] a means of determining whether a machine "thinks": Alan Turing proposed in 1950 that questions be fed to a machine and a human. If we cannot distinguish the machine's answers from the human's, then we must grant that the machine does indeed think. Thinking, after all, is just the manipulation of symbols, such as numbers or words, in order to achieve a certain goal.

Searle, who wanted to refute the idea of a strong AI, asks us to imagine a man who does not understand Chinese sitting in a room. The room contains a manual that tells the man how to respond to a string of Chinese characters with another string of characters. Someone outside the room slips a sheet of paper with Chinese characters on it under the door. The man finds the correct response in the manual, copies it onto a sheet of paper, and slips it back under the door. In this way, he mimics someone who understands Chinese, even though he himself does not understand a word. That's what computers do, too, according to Searle. They process symbols in ways that simulate human thinking, but they are actually mindless automatons. Searle's thought experiment has provoked numerous objections (Searle, 1983).

Let us look at one possible explanation. The Chinese room experiment is a splendid case of begging the question. The meta-question posed by the Chinese room experiment is: *How do we know whether any entity has a subjective, conscious experience?* When you ask this question, you are bumping into the solipsism problem. No conscious being has direct access to the conscious experience of any other conscious being. We cannot be absolutely sure that we ourselves or any other person is conscious, let alone a jellyfish, a robot, or a chatbot. We can only make inferences based on the behavior of that person, jellyfish, or robot. Delving deep into

[1] The Turing test is a method of inquiry in artificial intelligence (AI) for determining whether or not a computer is capable of thinking like a human being. The test is named after Alan Turing, the founder of the Turing test and an English computer scientist, cryptanalyst, mathematician, and theoretical biologist.

the issue of consciousness and conscious perception of the world, Daniel Denett's argument seems a relevant one: in his work, *From Bacteria to Bach and Back: The Evolution of Minds* (2012), Dennett elaborates on his long-standing claim that consciousness is overrated, at least when it comes to doing what we need to do to get through a typical day. We carry out most tasks with little or no conscious attention. Dennett calls this "competence without comprehension." Adding insult to injury, Dennett suggests that we are virtual "zombies," to which we could add "thorough believers." When philosophers refer to zombies, they mean not the clumsy, grunting flesh-eaters but creatures that walk and talk like sentient humans but lack inner awareness. Competence without comprehension may apply to menial tasks like brushing your teeth or driving a car or bike but certainly not to science and other lofty intellectual pursuits. But is this indeed so? Quantum physicists, for example, manipulate differential equations and matrices with impressive competence – enough to build quantum computers –but no real understanding of what the math means. "I think I can safely say that nobody really understands quantum mechanics," observed the physicist and Nobel laureate Richard Feynman. That's not surprising. Science makes progress by confronting our lack of understanding, and quantum mechanics has a reputation for being especially mysterious. So, if physicists end up like information-processing automatons, what hope is there for the rest of us? After all, our minds are habituation machines, designed to turn even complex tasks – like being a parent, husband, or teacher – into routines that we perform by rote, with minimal cognitive effort.

> *The Chinese room experiment could serve as a metaphor not only for physics but also for the human condition. Each of us sits alone within the cell of our subjective awareness. Now and then we receive cryptic messages from the outside world. Only dimly comprehending what we are doing, we compose responses, which we slip under the door. In this way, we manage to survive, even though we never really know what the hell is happening.*

4 Cognitive Biases Today

The following chapter describes an example, which illustrates a minefield of cognitive biases, based on our experience. People tend to pay attention to information about risks and are more inclined to share information about risks. We search for and remember things that fit well with what we already know and understand. These biases are products of our evolutionary past, and for tens of thousands of years, they served us well. People who behaved in accordance with them were more likely to survive than those who did not.

Modern technologies, however, are amplifying these biases in harmful ways. Search engines direct people to sites that fuel their suspicions, and social media connects them with like-minded people, feeding their fears. Making matters worse, bots – automated social media which are based on different AI agents and algorithms that impersonate humans – enable misguided or malevolent actors to take advantage of their vulnerabilities. Compounding the problem is the proliferation of online information. Viewing and producing blogs, videos, tweets, and other units of information called *memes*[2] have become so cheap and easy that the information marketplace is saturated (Dawkins, 2017). Unable to process all this material, we let our cognitive biases decide what we should pay attention to. These mental shortcuts influence which information we search for, comprehend, remember, and repeat to a harmful extent.

The need to understand these cognitive vulnerabilities and how algorithms use or manipulate them has become urgent. Researchers apply different approaches, such as cognitive experiments, simulations, data mining, and artificial intelligence, in order to comprehend the cognitive vulnerabilities of social media users. Analytical and machine learning aids are being developed to fight social media manipulation. Some of these tools are already being used by journalists, civil society organizations, and individuals to detect inauthentic actors, map the spread of false narratives, and foster news literacy. Let us have a quick look at some of the tools used for detecting inauthenticity.

4.1 *Information Overload*

The superabundance of information has generated intense competition for people's attention. As Nobel Prize-winning economist and psychologist Herbert A. Simon noted, "What information consumes is rather obvious: it consumes the attention of its recipients" (Newell & Simon, 1972). One of the first consequences of the so-called attention economy is the loss of high-quality information. Our social media newsfeeds are often so full that many of us can view only the top few items, from which we choose the reshare or retweet. Therefore, information overload alone can explain how fake news becomes viral.

This winner-takes-all popularity pattern of memes, in which most are barely noticed while a few spread widely, could not be explained by some of them being catchier or somehow more valuable. Virality resulted purely from the statistical consequences of information proliferation in a social network of agents with limited attention. Even when agents preferentially shared memes of higher quality, researchers observed little improvement in the overall quality of those shared the most. Even when we want to see and share high-quality information, our inability to view everything in our news feeds inevitably leads us to share things that are partly or completely untrue.

Cognitive biases greatly exacerbate the problem. We now know that our minds do this all the time: they adjust our understanding of new information so that it fits in

[2] *Meme* is an idea, behavior, or style that spreads by means of imitation from person to person within a culture and often carries symbolic meaning representing a particular phenomenon or theme. The word *meme* itself is a neologism coined by Richard Dawkins, originating from his 1976 book *The Selfish Gene*.

with what we already know. One consequence of this so-called confirmation bias is that people often seek out, recall, and understand information that best confirms what they already believe. This tendency is extremely difficult to correct. Experiments consistently show that even when people encounter balanced information containing views from differing perspectives, they tend to find supporting evidence for what they already believe. And when people with divergent beliefs about emotionally charged issues such as climate change are shown the same information on these topics, they become even more committed to their original viewpoints.

To make things worse, search engines and social media platforms provide personalized recommendations based on the vast amounts of data they have about users' past preferences. They prioritize information in our feeds that we are most likely to agree with and shield us from information that might change our minds. This makes us easy targets for polarization. But an analysis of consumption of low-quality information on social media shows that this vulnerability applies to both sides, in our case, to the believers and to the nonbelievers in ecological problems. Even our ability to detect online manipulation is affected by our ecological bias. As mentioned earlier, bots greatly reduce the quality of information in social networks. In real social networks, just a few early upvotes by bots can make a fake news item become viral.

4.2 Social Herding

In some situations, it may be better to run first and ask questions later (survival is humans' primary instinct). In the absence of clear signals, our brains use information about the crowd to infer appropriate actions, similar to the behavior of schooling fish and flocking birds. Such social conformity is pervasive. But when people are isolated into "social" groups, in which they can see the preferences of others in their circle but have no information about outsiders, the choices of individual groups rapidly diverge. Yet the preferences of "nonsocial" groups, where no one knew about others' choices, stay relatively stable. Put it differently, social groups create a pressure toward conformity so powerful that it can overcome individual preferences, and by amplifying random early differences, it can cause segregated groups to diverge to extremes (according to the fifth motivational law, *participation motivates*[3]). Social media follows a similar dynamic. We confuse popularity with quality and end up copying the behavior we observe. Various experiments indicate that information is transmitted via "complex contagion": when we are repeatedly exposed to an idea, typically from many sources, we are more likely to adopt and reshare it. This social

[3] Getting more people involved instead of less people involved can be the catalyst for the energy you need to get things done. Everyone who wants to be involved should be given the opportunity to make an impact, and often the peer pressure of watching others get involved can jump-start the others to get it in gear as well.

bias is further amplified by what psychologists call the "mere exposure" effect: when people are repeatedly exposed to the same stimuli, such as certain faces, they grow to like those stimuli more than those they have encountered less often.

Such biases are transformed into an irresistible urge to pay attention to information that is going viral – if everybody else is talking about it, it must be important. In addition to showing us items that conform with our views, social media platforms such as Facebook, Twitter, YouTube, and Instagram place popular content at the top of our screens and show us how many people have liked and shared something. Few of us realize that these cues do not provide independent assessments of quality.

As a matter of fact, programmers who design the algorithms for ranking memes on social media assume that the "wisdom of crowds" will quickly identify high-quality items; they use popularity as a proxy for quality. However, various independent analyses show that all platforms – social media, search engines, and news sites – preferentially serve up information from a narrow subset of popular sources. To understand why, we can observe how they combine signals for quality and popularity in their rankings. It can be concluded that agents with limited attention – those who see only a given number of items at the top of their news feeds – are also more likely to click on memes ranked higher by the platform. Another variable tracks the extent to which the ranking relies on popularity rather than quality. Even when we want to share the best information, the algorithms end up misleading us.

4.3 *Echo Chambers*

Most of us do not believe we follow the herd. But our confirmation bias leads us to follow others who are like us, a dynamic that is sometimes referred to as homophily, a tendency for like-minded people to connect with one another. Social media amplifies homophily by allowing users to alter their social network structures through following, unfriending, and so on. The result is that people become segregated into large, dense, and increasingly misinformed communities commonly described as echo chambers. This chambered structure efficiently spreads information within a community, while insulating that community from other groups. Even worse, social diffusion also makes negative information more "sticky." The information that had passed through people not only had become more negative but also was more resistant to updating.

4.4 *Rise of the Bots*

Information quality is further impaired by social bots, which can exploit all our cognitive loopholes. Bots are easy to create. Social media platforms provide the so-called application programming interfaces that make it fairly trivial for a single actor to set up and control thousands of bots. But amplifying a message, even with just a few early upvotes by bots on social media platforms, can have a huge impact on the subsequent popularity of a post. Bots also influence us by pretending to represent people from our in-group. A bot only has to follow, like, and retweet someone in an online community to quickly infiltrate it. Different research show that bots can effectively suppress the entire ecosystem's information quality by infiltrating only a small fraction of the network. Bots can also accelerate the formation of echo chambers by suggesting other inauthentic accounts to be followed, a technique known as creating "follow trains."

4.5 Curbing Online Manipulation

Understanding our cognitive biases and how algorithms and bots exploit them allows us to better guard against manipulation. There is a number of tools to help people understand their own vulnerabilities, as well as the weaknesses of social media platforms. One is a mobile app called *Fakey*[4] that helps users learn how to spot misinformation. The game simulates a social media news feed, showing actual articles from low- and high-credibility sources. Users must decide what they can or should not share and what to fact-check. Analysis of data from *Fakey* confirms the prevalence of online social herding: users are more likely to share low-credibility articles when they believe that many other people have shared them. Another program available to the public, called *Hoaxy*,[5] shows how any extant meme spreads through Twitter. In this visualization, nodes represent actual Twitter accounts, and links depict how retweets, quotes, mentions, and replies propagate the meme from account to account. These tools have been used by investigative journalists to uncover the roots of misinformation campaigns. Manipulation is getting harder to spot, however, as machine learning algorithms become better at imitating human behavior.

Apart from spreading fake news, misinformation campaigns can also divert attention from other, more serious problems. Programmatic tools are important aids, but institutional changes are also necessary to curb the proliferation of fake news. Education can help, although it is unlikely to encompass all the topics on which people are misled. Some governments and social media platforms are also trying to clamp down on online manipulation and fake news. But who decides what is fake or manipulative and what is not? Information can come with warning labels such as the ones Facebook and Twitter have started providing, but can the people who apply those labels be trusted? The dominance of social media platforms with global reach and close ties with governments further complicates the possibilities.

One of the solutions how to decrease manipulation and fake news in social medias may be to make it more difficult to create and share low-quality information.

---

[4] https://fakey.osome.iu.edu

[5] https://hoaxy.osome.iu.edu

We must be aware that free communication is not free. By decreasing the cost of information, we have decreased its value and invited its adulteration. To restore the health of our information ecosystem, we must understand the vulnerabilities of our overwhelmed minds and how the economics of information can be leveraged to protect us from being misled.

5 Scientific and Pseudoscientific Argumentation

The explosion of disinformation about global warming and other ecological problems could be called an *infodemic*,[6] which is the term used by the WHO to describe today's pandemic situation. An infodemic means too much information; however, its importance can also be passed on to environmental problems. It causes confusion and risk-taking behaviors that can harm the health of our planet. It also leads to mistrust in authorities and undermines the public response. An infodemic can intensify or lengthen outbreaks when people are unsure about what they need to do, for example, to protect our environment. With growing digitization – an expansion of social media and Internet use – information can spread more rapidly. This can help to fill information voids more quickly but can also amplify harmful messages.

Infodemic management is the systematic use of *risk- and evidence-based analysis* and approaches to manage the infodemic and reduce its impact on our environment. Infodemic management in relation to ecological problems focuses on enabling good practices through four types of activities:

- Listening to community concerns and questions.
- Promoting understanding of risk and health expert advice.
- Strengthening resilience to misinformation.
- Engaging and empowering communities to take positive action.

Infodemic management is widely known, but much less discussed is the role of *ostensible* "experts" or fringe scientists in perpetuating dangerous fictions. Since the dawn of the crisis, a disconcerting number of eminently qualified (pseudo)scientists and physicians have propagated falsehoods across social media, elevating themselves to the status of gurus in order to lend a veneer of seeming scientific legitimacy to empty, dangerous claims. And these bogus claims, like their pathological namesake, have gone uncontrollably viral. These fringe pseudoscientists are lauded as experts unafraid to speak truth to power. But it is crucial to note that these individuals, for all their formal credentials, extol a narrative completely at odds with reality, readily refuted by public bodies the world over. These pseudoscientific, conspiratorial claims are archetypal arguments from authority, where a perceived expert's

[6] An *infodemic* is too much information including false or misleading information in digital and physical environments during a disease outbreak. See: https://www.who.int/health-topics/infodemic#tab=tab_1

support is used to justify positions unsupported by data. Scientific claims do not derive their authority by virtue of coming from scientists but from the weight of the evidence behind them. Pseudoscience, by contrast, tends to focus on ostensible gurus (see *Kahneman system 1*) rather than a *consensus opinion* (see *Kahneman system 2*) (Kahneman, 2011). The only authority a scientist can ever truly invoke is a reflected one, dependent on accurately representing the evidence base. If they embrace fringe positions and jettison the principles of scientific skepticism, then their qualifications, education, and prestige mean absolutely nothing (Grimes, 2021).

> *Were these claims merely vapid that would be bad enough. Yet they are also uniquely damaging to public understanding. Real scientists and physicians must occupy an extremely trusted position in society, and an imprimatur of scientific legitimacy is a powerful one.*

To prove facts, scientists use complex scientific methods, which are largely incomprehensible to most people. Meanwhile, pseudoscientists use methods that are closer and more unambiguous to the everyman. Real science should be more approachable to the everyman; it should use such methods and arguments, which would be easier for people to understand, while at the same time not crossing the line of dismissing the accuracy and, above all, the reliability of what is being said, at the expense of improving intelligibility. If, while generalizing, we omit the less relevant and difficult-to-understand arguments and generalize to the level of the general population's understanding, then – as we reverse the generalization, adding more or less relevant data, which was omitted in the process of simplification – we should always be able to return to the starting point of the problem at hand, to the same complex scientific explanation.

The question arises, to what extent can we simplify and generalize (also, *who* can simplify and generalize)? In essence, the simplification process *must* be reversible, which means that a reverse process will bring us back to the starting point, which is symbolically shown in Fig. 2.3.

This is a trust utterly abused by fringe figures, who present qualifications as a proxy for scientific validity. This is superficially convincing to the point that it does not matter that simplified argumentation (e.g., different videos) originated in conspiratorial circles; the intrinsic aura of "science" afforded by apparent experts enables them to metastasize far beyond this odious origin. This in turn casts a specter of doubt over the advice of public bodies, distorting public understanding by presenting rank fictions in the stolen robes of science.

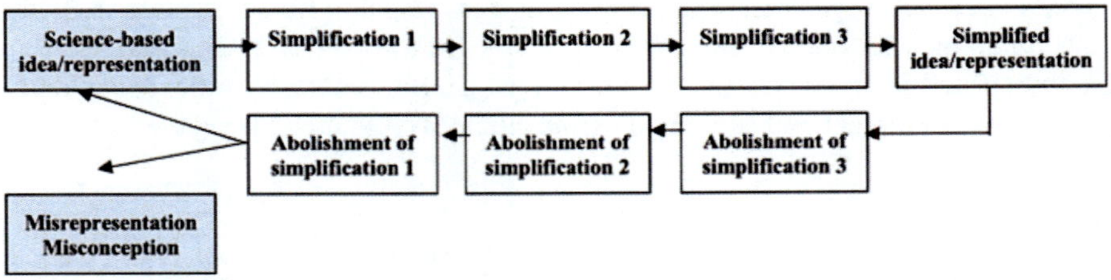

Fig. 2.3 The simplification principle. (Aberšek, 2012, 94)

The rise of pseudo-experts is perhaps symptomatic of a change in how we access information. As we become curators of our own media, the traditional gatekeepers and fact-checkers once implicit in most reporting have been increasingly sidelined. This in turn has made us more polarized and reduced our ability to differentiate fact from opinion. Motivated reasoning,[7] our human bias toward cherry-picking only arguments that chime with that we wish were true, most certainly plays a role. The impositions global warming and other ecological problems are manifold; it is not surprising that fringe scientists are inevitably invoked as sources for those with strong feelings against, for example, measures to reduce greenhouse emissions. Even if we are not ideologically predisposed to such positions, these claims undermine public understanding, blurring perceptions of scientific consensus and nudging us collectively toward fear and distrust.

The dark irony is that these fringe figures weaponize the societal trust afforded to science, unduly amplifying their capacity to unleash serious harm. To mitigate this, we need to keep in mind the vital distinction "science" and "scientists." Individual scientists are far from infallible; they can be fooled by subtle mistakes, haunted by spurious conclusions, or even become so ideologically wedded to a belief they bend facts to fit that preconception. Their motivations are human; they can be seduced by the lure of money, infamy, or admiration. Science, by contrast, is a systemic method of inquiry, where positions are formed on the totality of evidence. Crucially, to be labeled "scientific," ideas should be testable, and those that fail to withstand dispassionate investigation are duly discarded.

For all their qualifications, fringe scientists fail this basic tenet of science, as they are united in their willingness to embrace conspiracy theory when their claims are refuted. Lack of evidence for their position is airily dismissed as a cover-up by everyone from the authority on global level to the entire ecological establishment. But this performative outrage is so much sound and fury to distract from the inescapable reality that their positions are completely contradicted by the overwhelming weight of scientific evidence. This is scientifically reprehensible, and staggeringly irresponsible, conduct.

It is entirely understandable that many are left confused and uneasy by the vocal assertions of fringe figures, but the onus of proof is always on those making grand claims. The history of science and medicine is littered with the hubris of the arrogant and misguided, and mere credentials are no impediment to being wrong; only evidence truly matters. When confronted with the pronouncement of fringe figures, the motto of the Royal Society[8] should always be at the forefront of our mind: *Nullius in verba* (take nobody's word for it) (Grimes, 2021; Aberšek, 2021).

---

[7] *Motivated reasoning* is a phenomenon studied in cognitive science and social psychology that uses emotionally biased reasoning to produce justifications or make decisions that are most desired rather than those that accurately reflect the evidence, while still reducing cognitive dissonance.

[8] https://royalsociety.org/about-us/history/

6 Knowledge Is Power: Science, Technology, and Engineering Literacy

We live in a world that is, to a large extent, shaped by technology, so everyone needs to develop a certain attitude and adopt behaviors and awareness of technology. Our computers and our smartphones: technology. Our transportation: technology. Our homes, offices, and schools, our food and clothing, our heating and cooling, and our entertainment protection and pollution, energy production. They are all created and driven by technology. Likewise, many of the critical challenges that we face as a society, and that people will eventually need to address, have large technological components such as the quest to link experts throughout the world, the search for sustainable energy (Kutscher et al., 2018), dealing with global pandemics, and the development of environment protection and of environmentally friendly agriculture to feed a growing world population (NAGB, 2018; ITEA, 2003, 2007).

Despite its importance, however, technology has not been a focus of instruction and assessment in our educational system, particularly at the elementary and secondary levels. Some technologies are being integrated into the humanities, social sciences, science, and mathematics classes as methods for supporting learning in these subjects. Technologies are also becoming subjects of study, with the goal of developing an understanding of the technologies themselves, their various roles, and the engineering design processes to address human needs and wants but also human awareness. However, there are no standardized, national assessments to provide evidence of what technologically literate students know about science, technology, and engineering and the roles they play in our lives or the extent to which someone can understand and use technologies and understand how engineers design and develop them (Kordigel Aberšek & Aberšek, 2020).

Technology and engineering are increasingly being incorporated into school coursework, ranging from instruction on the use of computers and information technology within school subjects to classes that examine the role of technology in society or courses that teach engineering design. Information and communications technologies have become integral tools of the trade in academic, workplace, and practical contexts. Technology and engineering are essential components of contemporary science, technology, engineering, and mathematics (STEM) education (NAEP, 2013, 2014, 2018).

Nevertheless, in all contemporary and future intelligent systems and technologies, whether physical systems with a mind and body or just thinking algorithms in the cloud, the question of morals keeps coming back to us (Agar, 2012). Morals demonstrate the relationship of humanity to nature and society and are manifested as a sum of values (rules, norms, principles, categories, awareness, etc.), according to which we make decisions, what is good and what is bad, what is just and what is unjust, what is right and what is wrong, and in line with which we also behave.

Like it holds for all groups, society is also defined through the connections between its members, i.e., the individuals that constitute it. To understand society, we must research how it is affected by the physical environment, culture, and interpersonal relationships. Each of these spheres produces social values and awareness: e.g., education influences the economy and the attitude toward the environment, which changes the cultural patterns and awareness and thus affects the entire society (PISA, 2015). Moral considerations attempt to follow this when implementing different technologies into the society. The emphasis is on questions such as "are future entities going to have moral status?" (Bostrom, 2014), "who is responsible for the actions of technologies, and should we even create rules and an awareness that will have the status of a moral agent?" (Bryson, 2018), or "how to place technologies into the world so that it will act in accordance with our expectations?" (Rahwan, 2018), etc.

Ethics and technologies are thus connected on multiple levels:

1. *Ethics by design*: technical integration of ethical capabilities of reasoning.
2. *Ethics in design*: regulative and engineering methods that ensure the analysis and assessment of ethical consequences of the ecological systems' workings.
3. *Ethics for design*: codes of conduct, standards, and certification procedures that ensure the integrity of developers and users in researching, designing, using, and managing technological systems.

Therefore, we must make sure that the environmental models' reasoning will be based on abiding social norms and will be in accordance with basic human principles (morality) and that they will be able to justify their decision and awareness along with transparent functioning, with which they would ensure prosperity and progress in a sustainable world (Dignum, 2019).

Nevertheless, philosophy seems to have somehow forgotten certain important questions in the field of ethics of environmental solutions, questions that concern the morality and awareness in education, the most general being: "What are the moral problems in the usage of different kind of energy sources connected with STEM literacy?"

The technology and engineering literacy framework focuses on "literacy" as the level of knowledge and competencies needed by all citizens (PISA, 2015, 2018). People who are literate about technology and engineering are not expected to "do" engineering or produce technology in the professional sense. Therefore, the framework is not intended to address technical knowledge of specific technologies or types of engineering expertise taught in specialized courses that prepare students for postsecondary engineering studies. In the fourth grade, for example, all students can be expected to identify types of technologies in their world, design and test a simple model, explain how technologies can result in positive and negative effects, and use common technologies to achieve goals and awareness in school and in their everyday life. By twelfth grade, for example, all students can be expected to select and use a variety of tools and media to conduct research, evaluate how well a solution meets specified criteria, and develop a plan to address a complex global issue, also global ecological and environmental awareness.

7 Technology and Engineering Literacy

Any assessment of humans' technology and engineering literacy must start with a clear idea of exactly what technology and engineering literacy means (NAEP, 2013, 2014, 2018). That in turn requires clear definitions of technology and engineering. Historically the term "technology" has had a broader and deeper meaning, and it is this meaning that is represented in our case:

- *Technology is any modification of the natural world done to fulfill human needs or desires.* This definition sees technology as encompassing the entire human-made world, from the simplest artifacts, such as paper and pencil, to the most complex – buildings and cities, the electric power grid, satellites, and the Internet. Furthermore, technology is not just the things that people create. It includes the entire infrastructure needed to design, manufacture, operate, and repair technological artifacts, from corporate headquarters and engineering schools to manufacturing plants, media outlets, and distribution networks. Engineers are the agents for designing the many technologies that modify the world. Engineers may not actually construct artifacts such as calculators, bridges, or airplanes. They develop the plans and directions for how artifacts are to be constructed, as well as design processes such as assembly lines or procedures for green or not green environmental solutions.
- *Engineering is a systematic and often iterative approach to designing objects, processes, and systems to meet human needs and wants.* This framework defines technology and engineering literacy in a broad fashion.
- *Technology and engineering literacy is the capacity to use, understand, and evaluate technology as well as to understand technological principles and strategies needed to develop solutions and achieve goals.* Thus, technology and engineering literacy has much in common with scientific, mathematical, and language literacy. As with these other forms of academic literacy, technology and engineering literacy involves the mastery of a set of tools needed to participate intelligently and thoughtfully in society. The tools are different, but the ultimate goal is the same. One particularly important set of technological tools consists of information and communication technologies, and this framework sees knowledge of these particular technologies as an integral part of technology and engineering literacy (Kordigel Aberšek & Aberšek, 2020).

7.1 Areas of Technology and Engineering Literacy

The contents of technology and engineering literacy (NAEP, 2013, 2014, 2018) are organized into three major areas of technology and engineering literacy, each corresponding to a broad body of knowledge and skills with which students should be familiar:

- Technology and society
- Design and systems
- Information and communication

7.1.1 Technology and Society

This deals with the effects that technology has on society and the natural world and with the sorts of ethical questions that arise from those effects. Knowledge and capabilities in this area are crucial for understanding the issues surrounding the development and use of various technologies and for participating in decisions regarding their use.

7.1.2 Design and Systems

This covers the nature of technology, the engineering design process by which technologies are developed, and basic principles of dealing with everyday technologies, including maintenance and troubleshooting. An understanding of the design process is particularly valuable in assessing technologies, and it can also be applied in areas outside technology, since design is a broadly applicable skill.

7.1.3 Information and Communication

Technology includes computers and software learning tools, networking systems and protocols, handheld digital devices, and other technologies for accessing, creating, and communicating information and for facilitating creative expression. Although it is just one among several types of technologies, it has achieved a special prominence in technology and engineering literacy because familiarity and facility with it are essential in virtually every profession in modern society.

These three areas of technology and engineering literacy, as shown in Fig. 2.4, are closely interconnected. For example, to address an issue related to technology and

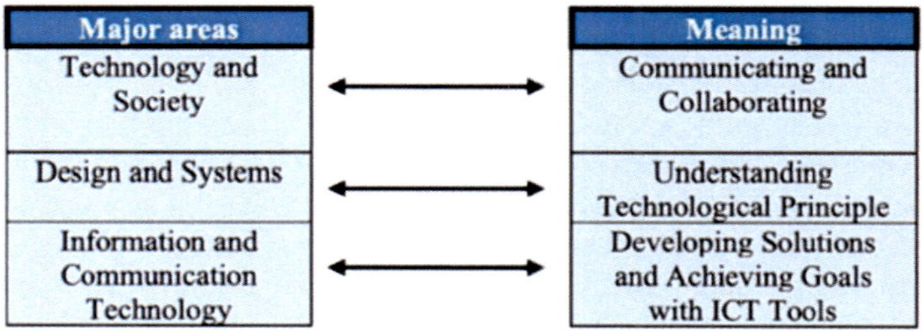

Fig. 2.4 Areas of technology and engineering literacy

society such as clean water, energy needs, digital-age communications, or global climate change, it is important to understand technological systems and the design process. Similarly, one must also be able to use various information and communication technologies to research problems, develop and communicate possible solutions, and achieve goals.

In all three areas of technology and engineering literacy, people are expected to be able to apply specific ways of thinking and reasoning when approaching a problem. These types of thinking and reasoning are referred to as practices, and the framework specifies three kinds in particular that people are expected to demonstrate when responding to test questions:

- Understanding technological principles, for example, what is green energy?
- Developing solutions and achieving goals
- Communicating and collaborating

7.1.4 Understanding Technological Principles

This focuses on how well people can make use of their knowledge about technology. This knowledge ranges from discrete declarative facts and concepts to higher-level reasoning about how facts, concepts, and principles are organized into various structures and relationships. People should be able to use their knowledge about technology to develop explanations and make predictions, comparisons, and evaluations.

7.1.5 Developing Solutions and Achieving Goals

This refers to peoples' systematic use of technological knowledge, tools, and skills to solve problems and achieve goals presented in realistic contexts. This practice includes both procedural and strategic capabilities – knowing how to apply simple steps and use technological tools to address real situations, as well as when and where to apply the tools and processes.

7.1.6 Communicating and Collaborating

This concerns how well students can use contemporary technologies to communicate for a variety of purposes and in a variety of ways, working individually or in teams or with peers and experts.

Regarding these three types of thinking and reasoning, it is necessary from the earliest age onward to develop an environmental or ecological awareness, i.e., it is necessary to foster an individual's attitude toward the environment. In the following section, we will focus more closely on the philosophy of mind and the related individual perception and interpretation of the world, which is the basis for the creation of an individual's awareness (Kordigel Aberšek & Aberšek, 2020).

8 Philosophy of Mind

Let us proceed from the premise that the philosophy of mind can be observed naturalistically. In doing so, the general schema can be understood as a *positivist requirement* (the mind can be adaptable) or as a *normative requirement* (the mind must be adaptable) (Kim, 1996). Based on this, we could define three borderline examples of how scientists in the past understood the world of nature and the problem of the mind in it:

- *A reductionist view*. This includes the "naturalists of the mind," whose main proponent was Galileo with his scientific method of resolution and composition (MRC).
- *A normative view*. Regulation of the relationship between the mind and the body, i.e., establishing rules or laws that would enable prediction and control. The main representative in science was Newton and in philosophy Locke.
- *The mind in biological terms*, sociobiology, Darwin's evolution theory.

With contemporary philosophers of the mind, two fundamental philosophical questions stand out:

- Whether mental phenomena such as feelings, intentional states (beliefs, desires, intentions, thinking, affirmation), and emotions can be reduced to interpretation using terminology and the language of the natural sciences.
- Whether mental phenomena correspond to the states and processes described by the natural sciences – physics, chemistry, biology, and neuroscience.

In recent history (second half of the twentieth century), when these questions came to the fore, there were quite a few interpretations and answers offered, but there has been no consensus in this area to this day. Let us summarize only a few of the suggested views:

- *Psychophysical reductionism (reductive physicalism)*, which reached its peak in the mid-1960s and then its decline in the early 1970s, argues that mental phenomena can be represented by physical facts and can be reduced to physical laws (Oppenheim & Putnam, 1958). Amplified metaphysical/ontological reduction supposes that psychological laws will eventually be explained in terms of the behavior of individual neurons in the brain; and that the behavior of molecules will eventually be explained in terms of the physics of atoms. If this is the case, then all psychological laws are in principle reduced to the laws of elementary particle physics. Of course, the following question arises: to what extent can we simplify (reduce) neuroscience so that we do not lose too much or stray too far from the essence of the problem? Even reductionists themselves do not agree on this matter.
- *Psychophysical antireductionism (non-reductive physicalism)*. Psychophysical antireductionism argues that while mental phenomena *depend* on physical ones, they are not identical with them, which is why the former cannot be reduced to the

latter. Thus, mental phenomena retain a genuine ontological status, while psychology remains an independent science with its own research methodology.

Representational theory is based on the approach according to which we represent mental states with propositional attitudes. The term propositional attitude, introduced by Bertrand Russell, refers to sentences where we use verbs in the main sentence, such as *believe*, *want*, *hope*, and *fear*, followed by a dependent sentence introduced by *that*. For example, *I wish that we leave behind a clean planet for the future generations*. Propositional attitudes have become the standard form for defining mental states. They allow us to have the same attitude toward different propositions or different attitudes toward the same proposition. In the computational-representational theory of thought, as developed by Fodor (1975, 1983, 1987), propositional attitudes are functional (computational) relations to mental representations. The fact that they are in an individual propositionally oriented state means that we have a representation of the content of the proposition adjusted in a functional role corresponding to this orientation. For example, to be thinking "It will rain. So I'm going inside" (Fodor, 1987, 17) is to express a mental representation, which means "I'm going inside," caused in a certain way, by expressing the mental representation that means "It will rain." The second statement of the representational theory of the mind speaks about the nature of mental processes: mental processes are causal sequences of expressions of mental representations (Markič, 2010).

8.1 An Epistemological Outline

Ancient Greek philosophers already asked themselves questions such as *How do I learn, remember, and perceive*? *What is the role of language*? *Does the soul survive the death of the body*? *What is the relationship between mental and physical processes*? *Are we free in the decisions we make*? *How do we know we know something*? Plato, Aristotle, Leucippus, Democritus, Epicurus, and Lucretius responded to such questions in various ways through their works, and later this line of thought continued with Descartes, Spinoza, Hume, Kant, and many others. In a short and very simplified manner, their attitudes toward the nature of the mind can be summarized with two sets of assumptions.

- *The first approach*. On the one hand, it seems to us that mentality is completely different from everything else in the world and must therefore have a special ontological status.
- *The second approach*. On the other hand, we accept that humans are part of nature and that the processes that take place inside human beings, including mental and physical processes, are also part of nature.

9 Recent Philosophical Views on Philosophy of Mind

If we want to understand the mind, the way in which humans think and reason according to the criteria of today's science, we need to focus on a view of science that recognizes and takes into account the general *explanatory gap* in psychology. What does this mean for the philosophy of mind? In a short and very simplified manner, attitudes toward the nature of the mind can be summarized with two sets of assumptions. On the one hand, it seems to us that mentality is completely different from everything else in the world and must therefore have a special ontological status. On the other hand, we accept that man is a biological being and a part of nature and that the processes that take place in him, including mental processes, are natural processes.

The first approach can be supported by the following assumptions:

- *The mind is dimensionless.* It seems to us that our feelings, desires, emotions, and thoughts have no spatial characteristics, while things around us have spatial characteristics such as size, shape, and position in space.
- *Direct access to mental states.* We come to know about our mental states and processes directly, through self-observation (introspection). When I feel a pain in my arm, that's enough to know I am in that condition. However, if we want to get to know the physical objects in our surroundings, we get to know them with the help of our senses, instruments, experiments, etc.
- *The problem of intentionality.* When we think, believe, or want, our thoughts, beliefs, and desires are always about something (e.g., I want us to go on vacation). They can also be about something that does not exist (e.g., about a round square). However, it is hard to believe that this would also apply to purely physical things.
- *The problem of consciousness*, for example, the conscious feeling of sadness overwhelming one as they read a sad story, or the occasional unbearable pain in the lower back – how could we analyze such feelings by means of scientific terms that are accessible to everyone?

The second approach can be supported by the following:

- If the processes inside humans are physical processes, then it is very likely that mental processes are also natural, physical processes.
- We have learned from the historical progress of science that the fact that we do not yet know how to explain a phenomenon by means of science itself – i.e., using the means and methods available to us at that time – does not mean that this phenomenon is something nonphysical or immaterial.
- From everyday experience, we can conclude that there is an interconnectedness between the mental and the physical.

While making assumptions based on the first approach, philosophers mostly derive from their everyday experience, while assumptions based on the second approach are mainly related to scientific findings. However, philosophical reflections have not been the main method of understanding mental processes for a long time.

At the turn of the nineteenth and twentieth centuries, when psychology became independent of philosophy following the example of empirical natural sciences, the leading German school of psychology emphasized introspection as a fundamental scientific method in psychology. Their basic hypothesis was that psychology is an exploration of the "phenomenology" of human mentality, so they sought to give a complete description of the map of mentality as presented to the subject. The problem with this approach was that the data collected on "sensory atoms" depended on the sensitivity and ability of the "introspectionists" so that they obtained different results, which undermined the "scientific nature" of the approach. Well, in addition to these "internal" causes, the decline was mainly due to a change in the research climate in Europe and the United States. A positivist attitude began to prevail in both science and the humanities. As a result, consciousness was pushed out of psychological research, as was the method of introspection, which was deemed unscientific and overburdened with questionable metaphysics. Behaviorism became established and with it the principle that psychology, as a scientific discipline, must study only what can be explored through observation, that is, observed behavior. The easiest way to imagine a behaviorist methodology is to imagine a human being as a "black box" that receives data from the environment and responds to it. The scientist is interested in the connection between stimuli and behavior, but he does not go into researching the "black box" itself, because according to him, everything that happens inside it is outside the level of scientific psychology. In this way, behaviorist psychologists avoided referring to both neurophysiological and mental states in their interpretation. However, it turned out that the project according to which all concepts with mental content could be defined with the help of operational definitions (as actual or potential reactions of the organism to certain stimuli) is unfeasible. Psychologists therefore turned to everyday commonsense psychology, which uses mental states (beliefs, desires, fears, and intentions) to explain behavior, and provided a framework that needed to be filled with scientific theory. This opened a new space for researching cognitive processes (perception, learning, concept formation, problem-solving, etc.) (Newell & Simon, 1972).

10 Ethics and Technological Development

Man is not adapted to life in nature like animals are, because in "raw" nature, he is unable to satisfy his basic needs (desires) in a natural way. In order to survive, people have to *know* nature and turn it into something useful and beneficial to ourselves. Only through work can people take from nature what they need, even that which is not in it according to the nature of things. Still, nature is the foundation of human existence. Human beings, who are inherently creatures of nature, become cultural beings through the processes of upbringing, education, and socialization, and because of this, our primordial, elementary ties with nature are weakened.

Because we are able to think and learn (and in turn, evolve), our knowledge and work are increasingly changing nature, from the so-called *natural environment* into a *new environment, the artificial environment.* People are able to satisfy their needs or desires more and more efficiently by using more and more efficient tools and more and more efficient devices to develop procedures that "facilitate working with the hands and make them stronger." From the tension between the unsatisfactory state of "the existing" and the strong desire for "something better," ever new problems arose, which people learned to solve by using some or other tool. At the moment in which such tools began coordinating between desires and actual needs, technology and engineering were born, or in other words, the tools that made life easier for people but which also gradually began to destroy the natural balance. The accumulation of experience and knowledge has created knowledge that man uses to his advantage and that is passed on to future generations.

Human work has always been associated with technology and technical aids. Immediately upon his creation, man became a creator of technology, as well as the user of his creation. With the industrial revolution, the development of technology took off like never before. Since then, technology has exploited and changed nature to such an extent that today we talk more and more about overexploitation, disturbances, and inappropriate interventions in natural processes and, unfortunately, about the catastrophic consequences of all this. Technology has shifted mountains and changed the course of rivers. But it has also destroyed forests; it has built but also demolished countless cities. Advances in technology have fulfilled more wishes than one could imagine even in fairy tales but also made it possible for even the worst nightmares to come true. The Shangri-La as a paradise where all of the work will be done by machines or as an apocalypse of atomic hell, technology can accomplish both (Lenk & Ropohl, 1989; Bonz & Ott, 2003). In developed industrial countries, nature is beyond being just tamed. The natural environment has completely changed there. However, when nature is damaged or destroyed, man's existence is also threatened, because he is, after all, a part of nature. And man is the only living being who consciously destroys the conditions for his further existence. Industrial societies are thus steadily evolving into societies with a high development risk, as the welfare society transforms into a multitude of individuals (Beck, 1992). Since the 1980s, development trends in chemical, atomic, and a number of other technologies have also become critical. The expansion of industry brings along the destruction of plants, animals, and man himself, as the pollution of soil, air, and water significantly reduces the quality of life.

Undoubtedly, the use of technics and technology has a negative impact on the natural environment, degrading, putting further pressure, and destroying it with each intervention. We must be aware, however, that technology is only a tool in the hands of people. We, the people, are the ones who can do good or harm with our thoughtful or thoughtless behavior: with considerate behavior, the negative impact could be much smaller than it is. Technology experts are familiar with the ways and means of putting minimum pressure on the natural and artificial environment, but decisions about using technology are rarely made by technology experts themselves; they are made by people who are most often interested only in capital and profit. However, they do not know much about technology itself and its impact on the environment. That is why it is necessary to think about the importance of technical competence, about engineering and technological literacy, and its implementation in those areas (and those social groups), which at first glance "have nothing to do with technology." From the very beginning, it is necessary to develop (in individuals and in the society as a whole) an awareness about the human impact (especially the negative aspect) on the environment.

"Humans are beings, who know how to, and are able to, take responsibility. It has been like this since time immemorial" (Lenk & Ropohl, 1989). To use technology means to use power. This is why one must always consider the use of technology from an ethical and moral point of view (Jones et al., 1989). We would not need ethics and morals if we were allowed to do everything we know how to. Moral judgment in the use of technology can be derived from the following criterion: is the use of technology going to benefit people (including the people using it), or is it primarily harmful? An ethical use of technology can be summed up by drawing an analogy to Kant's categorical imperative, "Act in such a way, so that the effect of your conduct will be in accordance with the permanence of human life on earth"; or expressed negatively, "act in such a way that the effects of your actions will not destroy future life chances"; or again positively, "in judging what you will do today, think of the future integrity of man and humanity" (Jones et al., 1989). When we talk about responsibility in the use of technology, we need to consider ethics and morality, as well as law. Morality is not as binding as law. The law is indeed based on ethics and morality, but ethics and morality do not have any tools (weapons, means of coercion) that could force us to comply with its norms. Ethics and morality have only symbolic, social controls in the form of reward and punishment. When we talk about a moral/ethical use of technology, we are talking about more responsible and mature judgment than when we justify our actions with legal norms, regulations, and standards.

The ethics of technology, or the ethics of using technology, includes ethical considerations and ethical reflection on the conditions, purposes, and consequences of developing, producing, using, and disposing of/recycling technology. In this context, it is worth highlighting the need for ethical reflection *before* the realization of individual technical solutions, as the consequences of unethical use of technology can be devastating to humanity (e.g., global warming). Technology itself is neither a moral nor an immoral force within society. Technology can become a moral force/power, or an immoral force/power, only after one makes the judgment on why and how to use it. In doing so, ethical reflection and awareness are crucial, as decisions based on technological and economic criteria are often not ethical.

Modern society, therefore, cannot avoid the issue of ethical judgment in the use of technology. For this reason, the competence of ethical reflection on the use of technology should be included among the key general objectives of any education system. At the level of the educational structure of lessons, professional engineering and technological knowledge and competences (problem solving, project management, and so on) are joined by *a value level*, according to which students learn about how to ethically reflect on the conditions in which technology can be used and on the goals we can achieve with it through the didactic process.

10.1 Morality, Machines, and Education

Nevertheless, the problems of implementing different advanced technologies into our lives are not a thing of the distant future or strictly a matter of academic discussions. Smart machines are already used in everyday life as an aid for accomplishing tasks more efficiently and for maintaining and even creating interpersonal relationships. This does not change when it comes to the field that is the most important for society, as it develops interpersonal relationships and culture and determines society in general. We are talking about the field of *education*, which is set with the relationship between the teachers and students, more precisely, with their behavior in the process of conveying and acquiring knowledge.

We are fairly good at simulating the external environment. However, everything becomes problematic when we attempt to do the same for internal environment, which is connected to the individual's *consciousness*. According to Chalmers (1996), we can split this problem into the following:

(a) *The easy problem*, i.e., the problem of how our conscious intentions cause the activation of neurons, which "do" what we planned.
(b) *The hard problem* (also known as the *explanatory gap*), i.e., the intentions which are the cause of our behaviors and consciousness, which brings us to the idea of free will, which is also connected to ethics or morality.

If we imagine every individual, including the student and the teacher, as a regulated system, then the educational system (which basically represents a social group), in which these individuals work, creates an external control, which assigns a certain value to the behavior of individuals. It is about prescribing social and moral norms in the given social group (which is generated mostly by the educational system) *and* in the entire society.

The hard problem is, as its name implies, even more demanding. Like it holds for all groups, society is also defined through the connections between its members, i.e., individuals that constitute it. In order to understand society, we must research how it is affected by the physical environment, culture, and interpersonal relationships, predominately because each of these things produces social values: e.g., education influences the economy and the attitude toward environment, which changes the cultural patters and thus affects the entire society. The moral considerations in the sphere of ecology attempt to follow this development when implementing technologies into the society.

11 Cognitive Science

How do we *think?* How do we *learn*, *remember*, and *dream*, how is *pleasure* created, where are *emotions* hidden, what is *free will*, how do we *make decisions*, what do we mean by *consciousness?* These are the kinds of questions, to which cognitive science tries to provide answers. Cognitive science addresses the sphere of human mentality

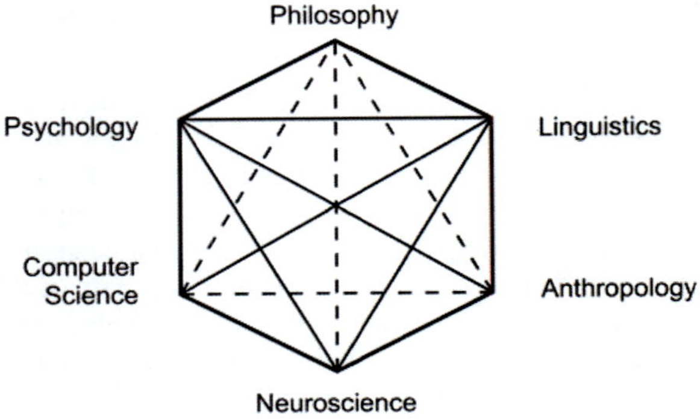

Fig. 2.5 Connections among the cognitive sciences

in an interdisciplinary manner – by connecting the findings of all disciplines that can say something about cognitive phenomena. Thus, cognitive science combines neuroscience, psychology, philosophy, linguistics, artificial intelligence, and the social sciences. It attempts to address mental processes holistically so as to arrive to a deeper understanding of the field that is closest to us in terms of experience (Aberšek et al., 2014a, b).

In recent decades, cognitive science has developed at a tremendous pace. Especially thanks to advancements made in neuroscience, we are slowly beginning to suspect that it is possible to scientifically undertake research into mental phenomena and even consciousness – a field of inquiry that was reserved, until relatively recently, only for mystics. Cognitive scientists are also trying to put their knowledge into practice – especially in the field of learning and teaching, collaborative work processes, and in the field of machine learning and decision-making.

Cognitive science grew out of the cybernetics movement in the 1950s and has undergone many paradigmatic changes since then. Cognitive science studies, which have flourished in all major world universities over the last 20 years, often lean against one of the constitutive disciplines (e.g., cognitive linguistics, cognitive neuroscience, cognitive anthropology, etc.), but it is becoming increasingly clear that only an equal treatment of all fields will provide a holistic coverage of human mental processes (Bermudez, 2010). The University of Berkeley presents the structure of cognitive science in the form of a hexagon, as shown in Fig. 2.5.

In the following section, we will address cognitive science mostly from the perspective of its philosophical foundations and on methods of cognitive learning.

11.1 Cognitive Learning Theory

Cognitive science could be defined most succinctly as a science that deals with the study of the mind, or human intelligence. Cognitive science is interdisciplinary and connects disciplines such as psychology, neuroscience, linguistics, philosophy, computer sciences (artificial intelligence), anthropology, biology, and physics into a comprehensive whole.

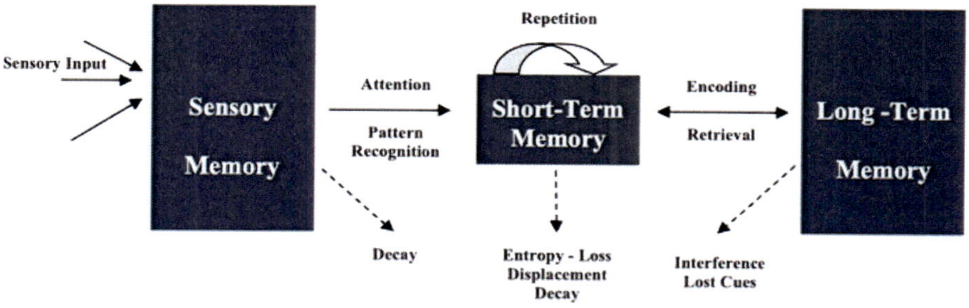

Fig. 2.6 Information processing model

Assumption: it is impossible to force someone to learn!

Cognitive learning theory defines learning as "semipermanent" changes in mental processes or associations. Cognitivists do not impose a requirement for an external demonstration of the learning process but rather focus on the internal processes and connections that are created during learning itself.

The basic premise of cognitivists (cognitive psychologists) is that there are cognitive processes which affect how we think and learn. The explanation of how cognitive processes work is known as the *information processing theory* (or *information processing model*). This simplified model, shown in Fig. 2.6, consists of three components of the information process and is the foundation of educational psychology.

In the last decades of the twentieth century, however, intensive development began in the field of cognitive modeling and intelligence studies, especially in connection with learning theories. J.R. Anderson became one of the pivotal figures in this field. In the 1970s, along with coworkers, he developed a complex theory of cognition based primarily on the ideas of Newell's and Simon's theory of symbols. Based on this theory, Anderson also developed a complex cognitive model of the human mind (brain) (Anderson, 2007). This model has developed intensively in recent decades and has led to a proper cognitive movement, which today includes hundreds of leading cognitive science experts from around the world.

12 Cognitive Learning

12.1 What Is Learning?

Learning is any process that modifies a system so as to improve, more or less irreversibly, its subsequent performance of the same task or of tasks drawn from the same population (H.A. Simon).

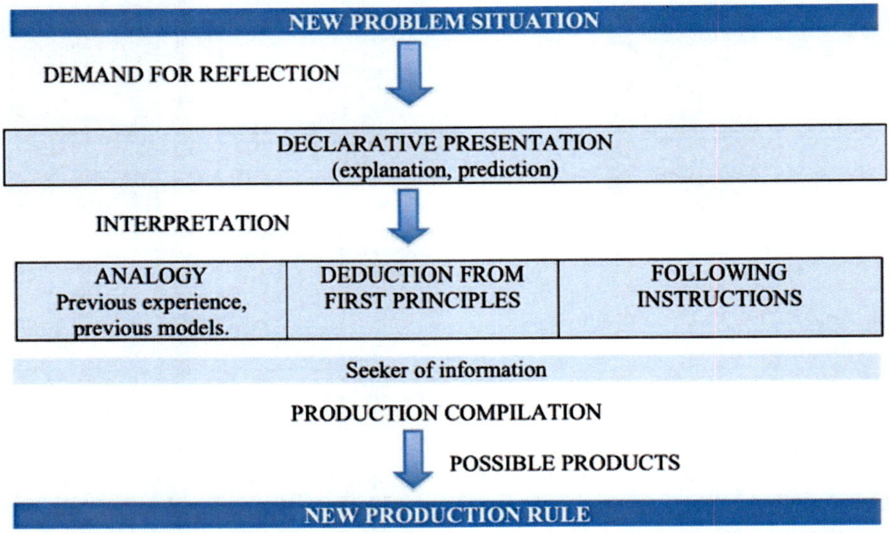

Fig. 2.7 Learning (cognitive domain)

Learning refers to the process that results in adaptive changes to the system. The changes that result from the learning process allow the same population to solve the same tasks faster and more successfully than before the learning process.

Learning could be defined by the following general situation: we have a system, a teacher who performs (wants to perform) a certain task. At first, the performance is bad. Through rehearsal, which consists of imitating a teacher and learning according to the trial-and-error principle, the performance gets better and better. "Better" can mean faster, more accurate, cheaper, etc., depending on the task. The rehearsal, the imitation of the teacher and repetition by correcting mistakes, is called *learning*.

Knowledge is defined as the interpretation of information in the form of data. Knowledge can be acquired either in advance or as a result of learning. It can be incorrect or correct but incomplete, etc. Any information with an explanation can be considered as knowledge. In practice, however, only useful knowledge is interesting, i.e., knowledge that enables the system to better solve tasks from a specific problem area, which is symbolically shown in Fig. 2.7.

12.2 What Is Upbringing?

When we talk about *upbringing*, we talk about social competence, emotional intelligence, and the closely related social awareness, i.e., about the human affective domain.

Awareness can be defined in a similar way as knowledge. Awareness is the acquisition of certain behavioral forms through the processes of learning, training,

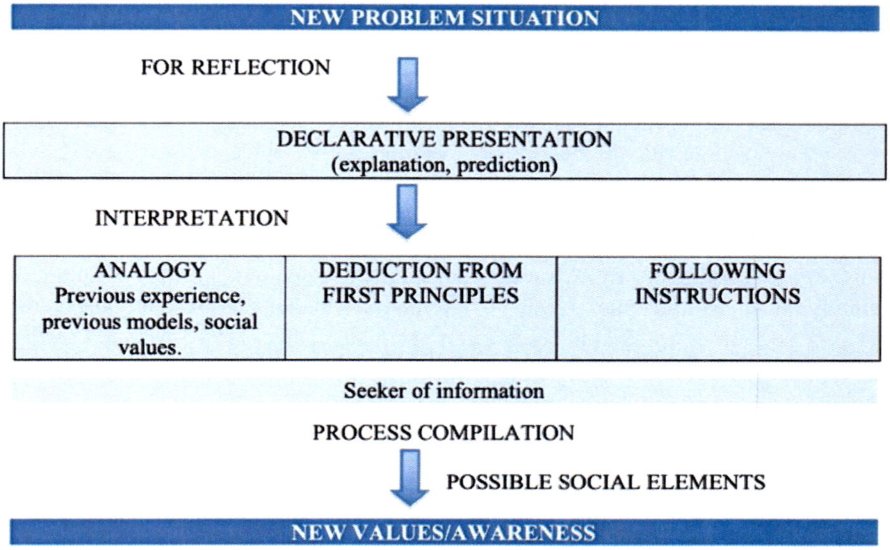

Fig. 2.8 Values, awareness (affective domain)

gaining experience, and interpreting information in the form of examples. Awareness can be acquired on an unconscious (e.g., by observation, by example) or conscious level, e.g., through learning. Like learning, awareness as part of education could also be symbolically depicted, as shown in Fig. 2.8.

Learning and awareness are present in all living beings and most obviously in humans. As a matter of fact, the entire natural evolution cycle could be described as learning: through genetic crossing, mutation, and natural selection, nature creates ever better systems capable of adapting to a wide variety of environments, the natural, the artificial, and the social.

12.3 Basic Forms of Learning for the Acquisition of Cognitive and Social Competence

1. **Controlled Learning**

Humans are the subjects of controlled learning in a sociological sense, that is, in the sense of upbringing and education, where the supervisor is an authority. Thus, our neuropsychological system only learns in a controlled way in a rather global sense when it comes to cognitive knowledge.

2. **Uncontrolled Learning**

Most learning in the brain takes place without control. In response to influences from the environment, the brain must find the optimal state by itself. In uncontrolled learning, parts of the (neural) network actually affect each other on their own, they coordinate and look for the best possible compromises, which are expressed as a stationary or equilibrium or stable state of the system.

3. **Reinforcement Learning**

This way of learning is a free form between controlled and uncontrolled learning. In reinforcement learning, the "teacher" only states whether the current state of the network is good or not. If it is not good, it needs to change. How it should change, however (unlike in controlled learning), is left to the network itself, i.e., to the network's random attempts. This category includes conditioning and reflex response (Pavlov) and various regulatory-vegetative functions of the nervous system.

12.4 The Impact of Learning on Social Competence and Emotional Intelligence

The development of a person's emotional intelligence is affected by two key factors, namely:

- The individual's primary social environment, e.g., family and family-related values.
- The individual's secondary environment, the social environment, as part of which school and education are vital.

When and if we are talking about any kind of *awareness*, in our case, an attitude toward a social issue, such as environmental issues, sustainable development, circular economy, ecological awareness, etc., we need to understand how such an awareness (consciousness) is developed in the individual. We will focus on this matter in more detail in the following section, using Bloom's taxonomy (affective domain) (Bloom et al., 1956) and the theory of emotional intelligence (Bohm, 2004; Goleman, 2006; Kahneman, 2011). But at this point, let us remain on a basic, philosophical level of interpretation. Let us look at this issue from the point of view of one possible (and fairly probable) philosophical viewpoint – naturalism.

13 Naturalism

The term "naturalism" has no precise meaning in contemporary philosophy. Its current usage derives from debates in America in the first half of the last century. The self-proclaimed "naturalists" from that period included John Dewey, Ernest

Nagel, Sidney Hook, and Roy Wood Sellars. These philosophers aimed to ally philosophy more closely with science. They urged that reality is exhausted by nature, containing nothing "supernatural," and that the scientific method should be used to investigate all areas of reality, including the "human spirit" (Krikorian, 1944; Kim, 2003).

13.1 Naturalization of Consciousness in Philosophy

We will need to ask ourselves about the essence of consciousness and about the essence of reacting consciously/subconsciously and creating relationships. What does it mean that objectivity "is" – this must become clear out of consciousness itself. However, since every consciousness is actually a consciousness "about something," this also includes the study of objectivity. We need to observe the ways of giving objectivity and the processes of explanation that follow it. By focusing on this relationship between objects and consciousness, a new science emerges – the phenomenology of consciousness. Instead of empirical consciousness (experiential), it has to do with pure consciousness (phenomenological). This goes to show that psychology should be more related to philosophy than it imagines.

The world of the psychological is fundamentally different from the known and established premises. The psychological essence is an experience through oneself or through remembrance. It is therefore always already being performed, involved in some stream of consciousness. As a phenomenon it has no reality, which is reminiscent of numbers. Phenomena are "in contact" with each other only through empathy; it is impossible to assign any kind of uniformness to them. In the psychological sphere, there is no difference between phenomenon and essence. The only way to repeat the same phenomenon is to remember, and only through remembrance can we find the being. Thus, the way of accessing phenomena is immanent vision.

The motto of naturalism, which we will try to defend in this work as well, is: "Man is a product of *race*, *environment*, and *the age* in which he lives" (i.e., *heredity*, *environment*, *time*).

Proceeding from this premise, we can arrive at a simple generalization. The fundamental human need throughout the course of history has been and still is the need for self-preservation, the need for survival. Of course, as humanity evolved, other, secondary needs have developed, such as the need for comfort, pleasure, etc. To satisfy especially people's primary needs, a nervous system has developed in living beings which:

1. Informs them of the needs of their *internal environment*.
2. Allows them to know what is happening in their *external environment*.

A human being perceives this external world through the senses. The brain then processes this sensory information to construct a certain representation of this world, drawing on memory to attribute pleasant or not-so-pleasant affective values to it. After that comes the behavior as such. It involves a movement whereby the human acts on its environment, so as to maintain its integrity as an organism.

Everything related to heredity and historical memory can be considered part of man's internal environment, while the naturalists' starting points, the environment and the time, can be considered as part of one's external environment. We have no direct influence on the internal environment, and we cannot change it in the "short term," but we can influence the external environment. As already mentioned, an individual's external environment can be influenced through the individual's primary social environment, i.e., family and family-related values, and through the individual's secondary environment, his social environment. The education system does not have a direct impact on an individual's primary environment, but it can and must affect an individual's secondary, social environment. We will focus on the latter in more detail below.

14 Developing Social Competence and Critical Thinking

In today's technologically developed world, we need to develop our (critical) thinking more and more quickly (today's information is worthless tomorrow) mainly on the basis of the following:

- Imperfect (poor, insufficient) information.
- Information that requires a huge amount of specific expertise, which in most cases is completely incomprehensible and often inaccessible.
- Misleading information (unintentionally and intentionally misleading).

As mentioned earlier, misleading information can be:

- *Unintentionally misleading*: many pseudoscientists and/or fringe scientists (politicians, influencers, etc.) must have their opinion even in the absence of relevant knowledge. Thus, they have to make decisions according to system 1 (Kahneman, 2011) quickly; however, because they do not even have the experience to make at least partly competent decisions, the probability of making wrong decisions (according to Murphy law[9]) is extremely high, incomparably higher than the probability of making correct decisions.
- *Intentionally misleading information* is provided by people who have specific interests and want to deliberately manipulate the audience. Such people do have enough correct information, but they distort the truth for reasons well known to them. Let us focus briefly on this, because such people are the most difficult to detect and they are also the most dangerous for every individual and thus for the whole society, as they are ready to destroy everything around them to achieve their partial interests.

[9] Murphy's *first law*: Anything that can go wrong will go wrong. . . . Murphy's *fourth law*: If there is a possibility of several things going wrong, the one that will cause the most damage will be the one to go wrong. Corollary: If there is a worse time for something to go wrong, it will happen then.

In the world of such false prophets (pseudoscientists), there is a huge number of people, let us call them believers (or, in modern terms, followers), who "blindly" believe them. They could also be called "true believers." *True believer* may refer to the following:

- One who strictly adheres to the tenets of a particular religious doctrine.
- By extension, one who is strongly attached to a particular *belief.*
- In the philosophy of Daniel Dennett, a system whose behavior is predictable via the intentional stance.
- In this context, it seems worth mentioning the *true believer syndrome*, a term for the irrational persistence of some untenable belief, for example, about nuclear power or global warming.

In the definition of the phrase true believer, certain terms such as doctrine, belief, true believer syndrome, and cognitive disorders require additional explanations:

- *Doctrine* (from Latin *doctrina*, meaning "teaching, instruction") is a codification of beliefs or a body of teachings or instructions, taught principles or positions, as the essence of teachings in a given branch of knowledge or in a belief system. Often the word *doctrine* specifically suggests a body of religious principles. *Doctrine* may also refer to a principle of law, in the common-law traditions, established through a history of past decisions. A *belief* is an attitude that something is the case or that some proposition about the world is true (the idea of something being true is discussed in Sect. 3: *How Do We Know, How Do We Believe?*). In epistemology, philosophers use the term "belief" to refer to attitudes about the world which can be either true or false. To believe something is to take it to be true; for instance, to believe that nuclear power is clean is comparable to accepting the truth of the proposition "nuclear power is clean." However, holding a belief does not require active introspection. For example, few carefully consider whether the sun will rise tomorrow (deductive assumption, see Sect. 5: *Scientific and Pseudoscientific Argumentation*), simply assuming that it will. Moreover, beliefs need not be *occurrent* (e.g., a person actively thinking "nuclear power is clean" or "global warming does not exist") but can instead be *dispositional* (e.g., a person who if asked about the nuclear power would assert "nuclear power is clean").

  There are various ways in which contemporary philosophers have tried to describe beliefs: beliefs as representations of ways that the world could be (Jerry Fodor), beliefs as interpretive schemes for making sense of someone's actions (Daniel Dennett), or beliefs as mental states that fill a particular function (Hilary Putnam). Some have also attempted to offer significant revisions to our notion of belief, including eliminativists, who argue that there is no phenomenon in the natural world which corresponds to our folk psychological concept of belief (Paul Churchland) and formal epistemologists, who aim to replace our bivalent notion of belief ("either we have a belief or we don't have a belief") with the more permissive, probabilistic notion of credence ("there is an entire spectrum of degrees of belief, not a simple dichotomy between belief and nonbelief").

Beliefs are the subject of various important philosophical debates. Notable examples include: *What is the rational way to revise one's beliefs when presented with various sorts of evidence*?; or *Is the content of our beliefs entirely determined by our mental states, or do the relevant facts have any bearing on our beliefs* (e.g., *if I believe that nuclear power is clean, is the nonmental fact that nuclear power is part of the content of that belief)?; How fine-grained or coarse-grained are our beliefs?*; and *Must it be possible for a belief to be expressible in language, or are there nonlinguistic beliefs?*

- The *true believer syndrome* is an informal or rhetorical term used by Lamar Keene and Spraggetti (1976) in book *The Psychic Mafia*. Keene used the term to refer to people who continued to believe in a paranormal event or phenomenon even after it had been proven to have been staged. Keene considered it to be a cognitive disorder and regarded it as being a key factor in the success of many psychic mediums.
- *Cognitive disorders (CDs)* are a category of mental health disorders that primarily affect cognitive abilities including learning, memory, perception, and problem solving. They are defined by deficits in cognitive ability that are acquired (as opposed to developmental), typically represent decline, and may have an underlying brain pathology. They define six key domains of cognitive function: executive function, learning and memory, perceptual-motor function, language, complex attention, and *social cognition*. Social cognition is a subtopic of *social psychology* that focuses on how people process, store, and apply information about other people and social situations. It focuses on the role that cognitive processes play in our social interactions. The way we think about others plays a major role in how we think, feel, and interact with the world around us. How exactly do psychologists define social cognition? While there is no single definition, there are some common factors that many experts have identified as being important. Social cognition involves the following:
  - The processes involved in perceiving other people and how we come to know about the people in the world around us.
  - The study of the mental processes that are involved in perceiving, remembering, thinking about, and attending to the other people in our social world.
  - The reasons we attend to certain information about the social world, how this information is stored in memory, and how it is then used to interact with other people.

Social cognition is not simply a topic within social psychology; it is an approach to studying any subject with social psychology. Using a social-cognitive perspective, researchers can study a wide range of topics including attitudes, person perception, prejudice, stereotypes, self-concept, discrimination, persuasion, *decision-making*, and other areas.

15 The Underlying Problem: The Environmental System

The *environmental system* may be understood in an ecological sense as the set of interactions between the elements of the *biosphere*, which includes the atmosphere, the hydrosphere, the lithosphere, and the ecosphere:

- The *atmosphere* is a mixture of nitrogen (78%), oxygen (21%), and traces (remaining 1%) of carbon dioxide, argon, water vapor, and other components. Although the atmosphere is approximately 1100 km high, the stratosphere (10–50 km) and the troposphere (less than 10 km) are the main atmospheric interactors of the biosphere. The atmosphere is a prime mean for the spatial diffusion of pollutants and a temporary mean of their accumulation until they precipitate.
- The *hydrosphere* is the accumulation of water in all its states (solid, liquid, and gas) and the elements dissolved it in (sodium, magnesium, calcium, chloride, and sulfate). Ninety-seven percent of the water forms the oceans, 2% is ice (north and south poles), and 1% forms rivers, lakes, groundwater, and atmospheric vapor. It covers around 71% of the Earth's surface and is an important accumulator of pollutants and a significant vector of diffusion.
- The *lithosphere* is the thin crust between the mantle and the atmosphere. Although the lithosphere is around 100 km thick, only 1 km of it can be considered in interaction with the biosphere (notably through aquifers). Main components are oxygen (47%), silicon (28%), aluminum (8%), iron (5%), calcium (4%), sodium (3%), potassium (3%), and magnesium (2%) in a crystalline state. The lithosphere is the main source of pollutants since it is from where raw materials are extracted to support life and economic activities and a permanent accumulator. Some pollutants are naturally released through sources like volcanic eruptions, while others like fossil fuels are the result of artificial extraction and combustion.
- The *ecosphere* is the set of all living organisms, including animals and vegetation. They are temporary accumulators (like lead) and sources for pollutants (natural forest burning) in a very complex set of relationships with the atmosphere, hydrosphere, and lithosphere (Fig. 2.9).

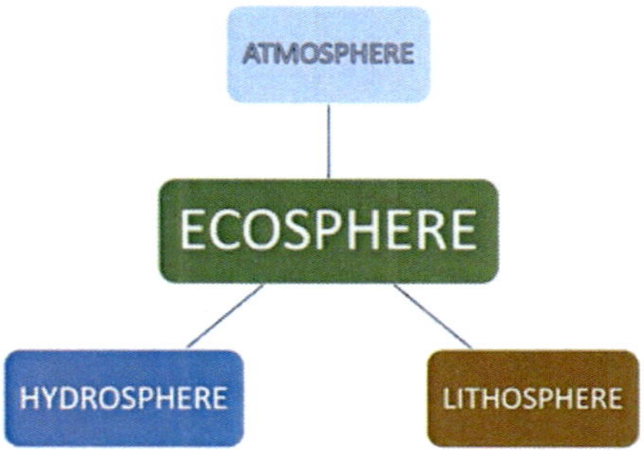

Fig. 2.9 The environmental system

The *environmental dimensions* sequentially involve the following:

- *Causes*. Two major factors are contributing to the level of different activities. Economics refers to the general level of development and income. An advanced economy is likely to generate more activities per capita than a developing economy. Land use refers to the spatial structure and location of the activities demand, which indirectly influences the modes used to support spatial interactions generated by economic activities.
- *Activities*. Involve a wide array of factors expressing the usage of infrastructures and all the related services to support the production system, particularly operations. All these activities have environmental outputs.
- *Outputs*. The first outcome of different activities is emissions of all sorts (carbon monoxide, nitrogen oxides, particulates, etc.). According to the geographical characteristics of the area where emissions occur (e.g., wind patterns), ambient pollution levels are created. Once these levels are correlated with population density and activity levels, a level of exposure to harmful pollutants can be calculated. This exposure is likely to have consequences.
- *End results*. They include all the health, environmental, and welfare effects of the exposure to emissions from different activities, which are very difficult to measure (Fig. 2.10).

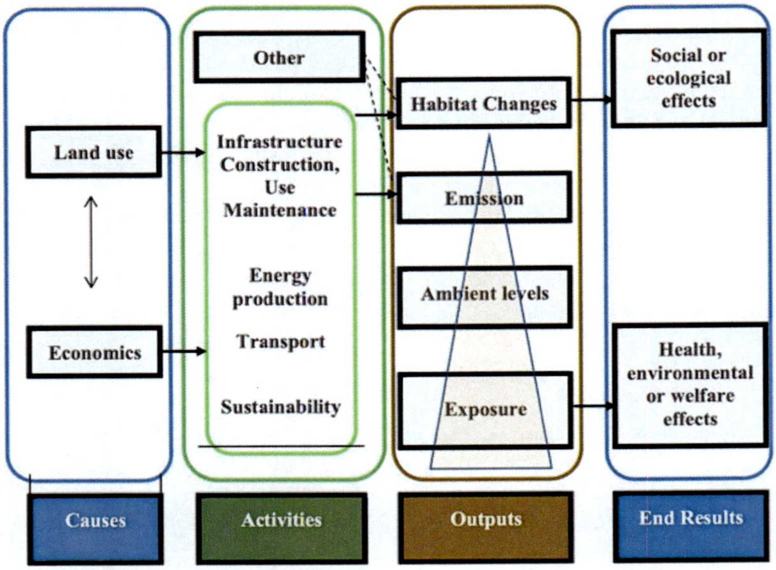

Fig. 2.10 Environmental dimensions of industrial production

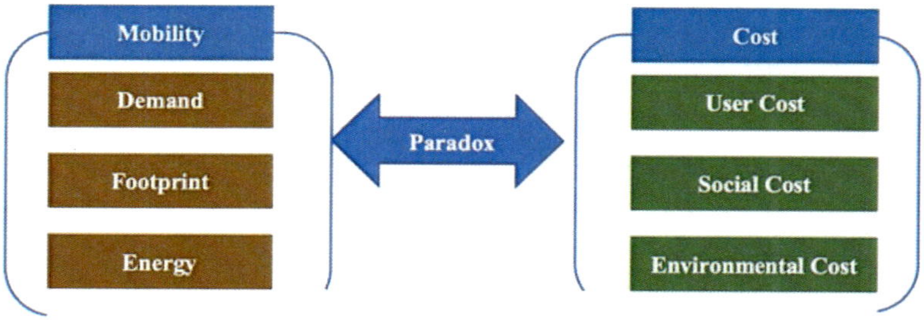

Fig. 2.11 The paradox of mobility and its costs

The issue of all kind production and the environment is *paradoxical* since industrial production conveys substantial socioeconomic benefits but at the same time impacting the whole *environmental systems*. From one side, different activities support increasing mobility demands for passengers and freight, while on the other, this activity is associated with *environmental impacts and environmental footprint*. Further, environmental conditions impact production systems in terms of operating conditions and infrastructure requirements such as construction and maintenance (Fig. 2.11).

The growth of production has expanded the source of emission of pollutants and their multiple *impacts on the environment*. These impacts fall within three categories:

- *Direct impacts*. They are the immediate consequence of different activities on the environment where the cause-and-effect relationship are generally clear and well understood. For instance, noise and carbon monoxide emissions are known to have direct harmful effects.
- *Indirect impacts*. They are the secondary (or tertiary) effects of different activities on environmental systems. They are often of a higher consequence than direct impacts, but the involved relationships are often misunderstood and more challenging to establish. For instance, particulates, which are mostly the outcome of incomplete combustion in an internal combustion engine, are indirectly linked with respiratory and cardiovascular problems since they contribute, among other factors, to such conditions.
- *Cumulative impacts*. They consider the varied effects on ecological footprint, of direct and indirect impacts on an ecosystem, which are often unpredictable. Climate change, with complex causes and consequences, is the cumulative impact of several natural and anthropogenic factors, in which transportation plays a role.

The complexities of the impacts have led to much controversy in environmental policy (UN-HABITAT, 2009). This is made even more complex by the fact that priorities between environmental and economic considerations shift in time, which can have an impact on public policy. Public preferences concerning the environment have shifted in the last 30 years. Although environmental matters were high in the

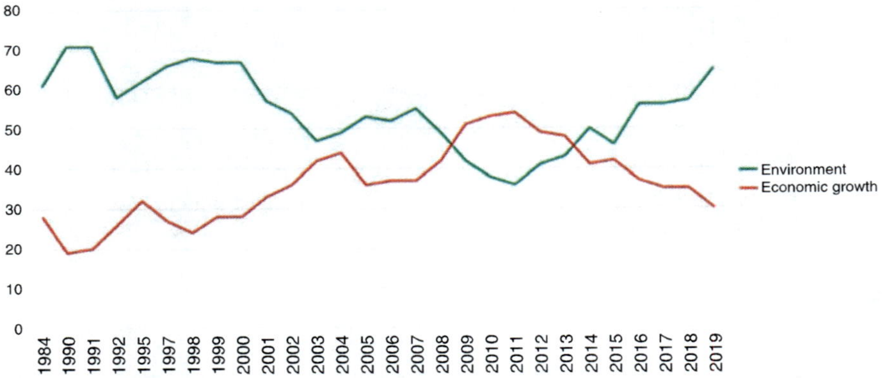

Fig. 2.12 Public Preferences for Priority between the Economy and the Environment, 1984–2019

public agenda in the 1980s and 1990s, they receded in the 2000s. There are several factors at play behind these fluctuations. First, environmental concerns tend to shift with economic conditions. When economic prospects are good, environmental concerns are more prevalent, while when economic prospects are more limited, they are less prevalent. For instance, the recession of 2001, the financial crisis of 2009, and the pandemic of 2019/2020 were all linked with sharp drops in environmental concerns (Fig. 2.12).

Another factor that can explain the decline of environmental concerns relates to the regulation and mitigation of the most visible environmental issues in the 1970s and 1980s, which resulted in improvements. This is particularly the case for air emission standards, such as vehicles that became less polluting and more energy efficient. Many institutions were created to regulate environmental matters for the sake of the public. Since 2010, environmental concerns are on the rise, mostly attributed to a growing awareness about climate change and its potential impacts on human activities.

Total costs incurred, for example, with energy production, notably environmental damage, are generally not fully assumed by the users. The lack of consideration of *real costs of the energy* could explain several environmental problems. Yet, a complex *hierarchy of costs*[10] is involved, ranging from internal (mostly operations), compliance (abiding by regulations), contingent (risk of an event such as a spill), to external (assumed by the society).

[10]The *cost hierarchy* is a classification system used in activity-based *costing* that designates activities based on how easily they can be traced to a product. These involve activities targeted at a specific product or product line, such as the *cost* to process an engineering change order.

16 Conclusion

How do people think and perceive truth(s)? How do we respond to these truths? What kind of knowledge is necessary for us to understand problems? These are the basic questions, to which we attempt to provide answers in the following chapter.

Scientists (experts) often communicate to us things about the world, which we would otherwise never think about, consider, or believe. *But how do scientists arrive to such very amazing realizations?* The answer, again, appears simple at first sight:

By thinking and reasoning!

Once we have arrived, in some way or another, from reasoning to certain conclusions, we need to confirm (or reject) these conclusions; we need to provide argumentation. Argumentation tries to take effect on an audience, change the opinion or atmosphere of the audience, and reach a consensus (a compromise through conversation) and not to enforce an own will by means of coercion or dressage. In science, this is where we use the old maxim: "use the power of argument rather than arguments of power." However, it is very often that we stumble upon a basic problem. Our knowledge and our perceptions are something that we experience through our perceptual world. But one may ask oneself whether we can truly believe ourselves and truly rely on our perceptions, our own mental images. To prove this, philosophers use various tools such as thought experiments. To illustrate our own mental images, let us consider John Searle's *Chinese room experiment.*

The Chinese room experiment can serve as a metaphor for human understanding. Each of us sits alone within the cell of our subjective awareness. Now and then we receive cryptic messages from the outside world. Only dimly comprehending what we are doing, we compose responses, which we slip under the door. In this way, we manage to survive, even though we never really know what the hell is happening. Speaking about the environment, we are faced with the same fundamental problem: we do not really know; we are not aware of how our environmental system really works.

Chapter 3
Environmental Costs Hierarchy

Abstract The third chapter, entitled *Environmental Costs Hierarchy*, deals with the human impact on the environment, especially in terms of the ecological footprint, and the related concept of human ecological awareness. Environmental costs are costs connected with the actual or potential deterioration of natural assets due to economic activities. They can be observed from two different perspectives, namely, as *costs caused*, that is, costs associated with economic units actually or potentially causing environmental deterioration by their own activities, or as *costs borne*, that is, costs incurred by economic units independently of whether they have actually caused the environmental impacts.

Environmental costs are *costs connected with the actual or potential deterioration of natural assets due to economic activities*. Such costs can be observed from two different perspectives, namely:

- *Costs caused*, that is, costs associated with economic units actually or potentially causing environmental deterioration by their own activities.
- *Costs borne*, that is, costs incurred by economic units independently of whether they have actually caused the environmental impacts.[1]

Cost type and implication is defined below.

Cost type	Implications
Internal	Material, labor, other expenses, and revenues that are commonly allocated to a product or process. Can easily be quantified (internalized)
Compliance	Expenses incurred by and benefits to the firm that are not related to products or processes. Mostly concern compliance to regulations
Contingent	Potential liability or benefit that depends on the occurrence of a future event. Assessed as a risk

(continued)

[1] https://stats.oecd.org/glossary/detail.asp?ID=819

B. Aberšek, A. Flogie, *Human Awareness, Energy and Environmental Attitudes*,
https://doi.org/10.1007/978-3-031-05871-4_3

Cost type	Implications
Image/ relationship	Costs/benefits related to the subjective perceptions of a firm's stakeholders
External	Costs/benefits of a company's impacts upon the environment and society that do not directly accrue to the business. Difficult to quantify (externalized)

Source: Adapted from the US Environmental Protection Agency (2000) The Lean and Green Supply Chain: A Practical Guide for Materials Managers and Supply Chain Managers to Reduce Costs and Improve Environmental Performance, Environmental Accounting Project

1 Ecological Footprint

A very short definition of *ecological footprint* would be to define it as the impact of human activities measured in terms of the area of biologically productive land and water required to produce the goods consumed and to assimilate the wastes generated (Fig. 3.1).

Since the first calculation in 1961, the global ecological footprint has increased from 1.7 gha/person to 2.75 gha/person in 2016. In 2020, it increased to over 8 gha/ person with the largest polluters. Assuming that the planet's biological capacity is 1.65 gha/person (12.2 billion in total), the global ecological footprint in 2016 exceeded the global biocapacity by 1.1 gha/person or by about one-half of our planet's biological capacity.

Therefore, we are talking about an ecological deficit, which has been increasing since the 1970s. Increasing the ecological deficit means that natural capital is consumed faster than it is renewed, which in the long run represents a reduction of the planet's biocapacity.

We use the ecological footprint to estimate how many planets equal to Earth would be needed for human survival. In 2016, the footprint of all people in the world

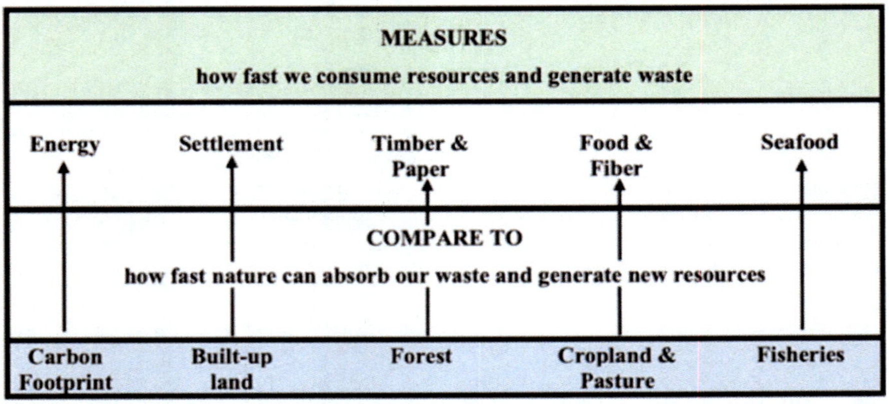

Fig. 3.1 The ecological footprint

was estimated at 1.5 planets, which means that we consume natural resources 1.5 times faster than they are renewed; in 2020 the number rose to 1.7 planets. It is estimated that by 2030 we will need two planets. The key message to the world could be formulated thus:

We are *not keeping the ecological footprint within acceptable limits*, which causes an imbalance between the *demand* (for food, fuel, wood, and fiber) and the *biological capacity for self-renewal.* Demand is almost twice the biological capacity. The greatest pressures on the environment are caused using nonrenewable energy sources (fossil fuels), especially in the energy sector and in transport. The fundamental problem in this area is that many individuals and consequently a large part of society are not aware that this problem exists, and what is even more concerning is that a large part of society is not in fact unaware of the problem but rather that it deliberately belittles the problem and partially or completely denies it. It does not acknowledge the existence of this problem and does not believe the environmental protection experts that this problem is indeed a burning issue. From a philosophical point of view, we could ask ourselves a simple question, *what is true and what is not?* And since knowledge alone, the mere awareness of the problem itself is not enough, it is necessary to find a solution to this problem (as well as to any such problem). The solution, however, is a vicious circle, as the solution lies only in:

> *...being aware of the problem, realizing that this is our problem, realizing that we too can be a part of the solution to this problem, and making a conscious decision that we must respect the solution to the problem, even if we sometimes have to give up something we like.*

We need to start building this kind of awareness in the individual as soon as possible, in young children and then throughout the entire life span of the individual. Two words are key in this process: *upbringing and education.* But how do we acquire knowledge, skills, and awareness? How do we develop *the affective domain*, *our emotional intelligence*, and *social competence*, which are key to solving this problem? Let us articulate this through the learning process and the example of *cognitive learning theory*, one of the most modern approaches related to both cognitive science and neuroscience, which is symbolically shown in Fig. 3.2.

2 The Human Population and Human Awareness

As we concluded earlier, the solution to environmental issues is based on awareness and in turn on appropriate actions of each individual and consequently the development of the whole society at a global level. Everyone must act within their means: individuals need to develop an appropriate attitude toward the natural environment, and the whole society (politics, public institutions) needs to create suitable conditions, appropriate attitudes, and supporting legislation.

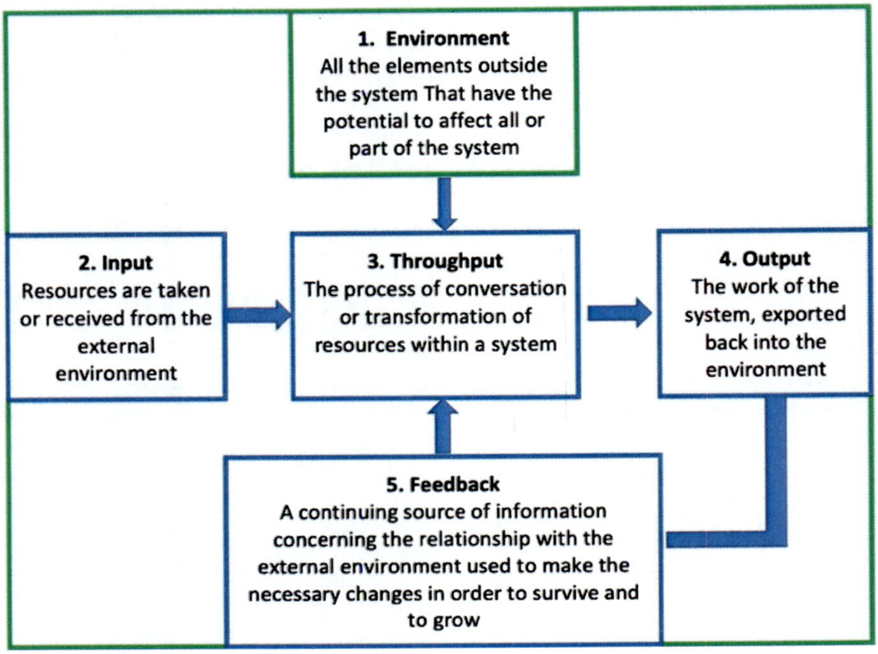

Fig. 3.2 A systematic approach to solving environmental issues

The advancement of mankind, its development and science, are best characterized by the following two determinants:

- *The exponential rate of the development* of society and consequently all its component parts.
- *The technological tone* of all components of this development. Technologies shape the society of today and will continue to shape, at an ever-increasing pace, the society of tomorrow. By the term "technology," we understand the following: "Technology means two things, a cause and an effect for all rapidly changing societies. When we talk about technology today, we talk mostly about intelligent systems, about the so-called cybernetic systems, which include *hardware* – machines and devices (or, the human body) and *software* – the intelligence, the mind, the human *feelings, intentional states, emotions, awareness,* etc. Many problems related to modern technologies are also related to the notion of artificial intelligence. For solving environmental issues in relation to AI, we require a general knowledge of the world, society, and people, which makes this issue a *philosophical problem*. It is thus extremely important for all decision-makers in the wider field of ecology to *understand* the basics of the general mechanisms of intelligence as part of the philosophy of mind, i.e., how nature works and how human intelligence works" (Aberšek, 2013).

Resulting from this is the certainty that only those parts of society – or those parts of technology – that will be able to give way to these two trends will be able to survive. We do not need to turn a blind eye to the fact that our society is trapped in an ever-faster spiral of development, which, in the final consequence, tends toward something that will be harder and harder to maintain from today's point of view – regarding society and the social changes within it. A logical question can be asked here: *Are we at the verge of collapse, the point of no return, on the edge of chaos?* But what is chaos, not in the general sense of the word but as a scientific term? The chaos theory studies the reactions of dynamic systems (a school system, e.g., and the Earth's natural environment are both dynamic systems), which are very susceptible to initial conditions because small changes in these conditions can lead to massive changes in the way the system and its component parts work. In short, when a system is close to the state of chaos, the possibility to introduce change is at its highest. We, however, need to realize that these changes can be good-correct or bad-incorrect. As Edward Lorenz said, "Chaos is when the present determines the future, but the approximate present does not approximately determine the future." In other words, in the state of chaos, even the most sophisticated methods of predicting the future cannot give us adequate results.

Let us confirm the fact that we are coming closer to the edge of chaos by looking at two examples:

1. The increase in population throughout history. A statistic is infallible, and, based on it, we can ask a simple question: *How many people can our planet sustain?*
2. What effects do the increase in population, the technological adjustments, and the energy needs of the natural environment (ecological footprint) for human comfort have on the health of our planet?

In both cases, we are talking about the ecological footprint. A great deal has already been written about population growth; the problem can be illustrated by using two images. Table 3.1 shows the number of people on our planet and the number of years when the population increased by one billion. At the beginning of

Table 3.1 Population growth

Year	No. of inhabitants (in billions)	No. of years for increasing population for 1 billion
1804	1.0	–
1927	2.0	123
1959	3.0	32
1974	4.0	15
1987	5.0	13
1998	6.0	11
2011	7.0	13
2016	7.4	–
2019	7.65	–
2021	7.9	–
2025	8	14

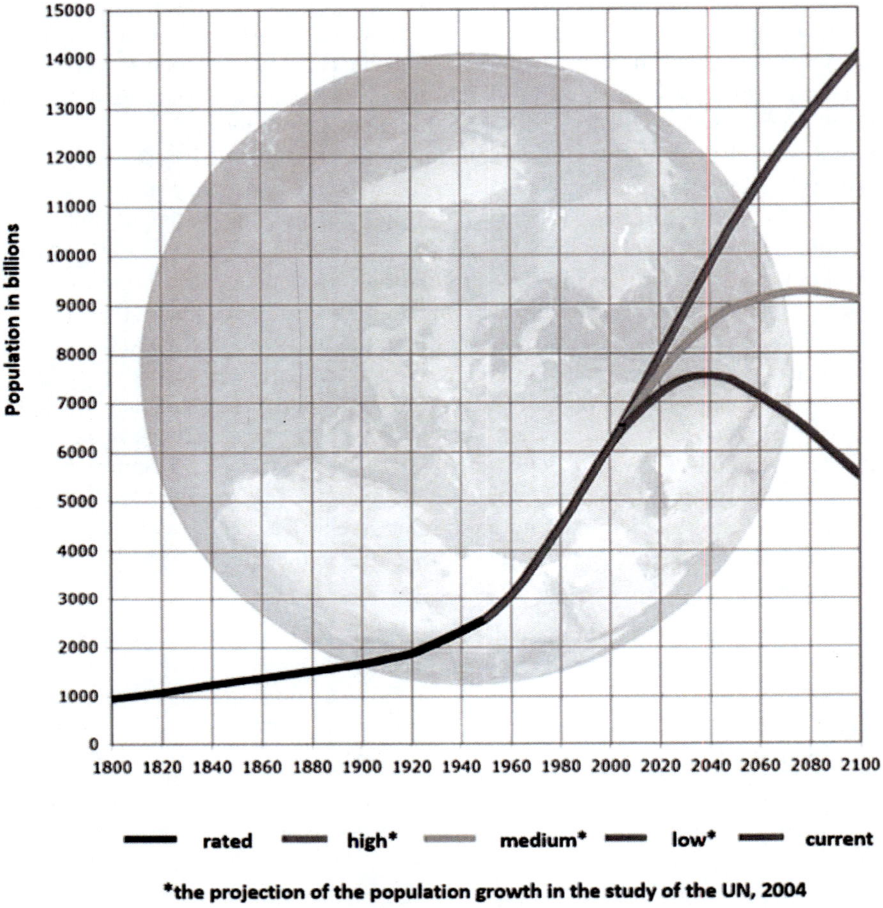

Fig. 3.3 Population growth projection

the twentieth century, we would have needed more than 100 years to achieve this number. Then, in barely 100 years, this number was reduced by a factor of 10, i.e., it appears now that we will be reaching one billion in just a little over a decade. This data is self-explanatory. At the same time, the prognosis for the next century does not look promising, as shown in Fig. 3.3, which shows an estimated growth of population until the beginning of the twenty-second century. Although the predictions are somewhat contradicting, even according to the average prediction, the outcome does not look promising.

As we mentioned at the beginning of this book, there are only three major monopolies in the world today, *food*, *technology*, and *energy*. In the last century, our needs have skyrocketed.

One does not need to be a scientist to arrive to certain obvious conclusions from what has been said. From the diagram above, it is evident that our growth ratio is daunting. We can see that in the last 50 years, our crimes against nature grew exponentially, and we can only ask ourselves, what will come sooner? Will we run out of resources, or will we poison ourselves first? Thus, *the question of whether we are at the edge of chaos is in fact no longer a question.*

Based on the above, we can make two conclusions:

- Being at the edge of chaos can also be good news, because it gives us ample opportunities to change the trend of development (just think of the pandemic – we developed vaccines in 1 year's time!); the only problem left is that *we have to know what we want.*
- We only have a small amount of time to do it; we do not have the luxury of learning from our mistakes. *We cannot afford to make any more mistakes!*

If all of this is true for the entire system, then it is also true for its component parts, the state of the subsystem, the social system (i.e., humans), which is certainly the main reason for this chaotic state. At this point, we will only focus on a small part of the social system, on *education*, and on the question of how we should develop a *sociological and anthropological awareness related to our environment.*

Though people live in the here and now, they have the capacity to think about future events, remote places, distant others, and alternative realities. This ability to think about events that extend beyond the immediate context is one that many might take for granted. There is some research that suggests, however, that this may be a uniquely human skill. *Construal level theory (CLT)* attempts to explain how people accomplish this remarkable feat. Central to understanding CLT is the idea of *psychological distance*. The qualifier "psychological" in the term "psychological distance" refers to the idea that the mind treats these different dimensions of distance in a mentally similar manner (Trope & Liberman, 2010).

CLT is a theory in social psychology that describes the relation between psychological distance and the extent to which people's thinking (e.g., about objects and events) is *abstract* or *concrete*. The general idea is that the more distant an object is from the individual, the more abstract it will be thought of, while the closer the object is, the more concretely it will be thought of.

According to CLT, people perceive events that vary in several types of psychological distance:

- Temporal distance (time).
- Spatial distances (physical space).
- Social distances (interpersonal distances, such as distance between two different groups or two dissimilar people).
- Hypothetical distances (imagining that an event is likely or unlikely).

Psychological distance (see also the affective domain of Bloom's taxonomy) affects the extent to which we think about an event, person, or idea as high or low level, and this will influence how concrete or abstract those thoughts are:

- *High-level construal is when people think abstractly.*
- *Low-level construal is when people think more concretely and is associated with psychological proximity.*

3 Construal Level Theory and the Theory of Chaos

In *a state of chaos* (in the short term, according to the principle that the closer the object is, the more concretely it will be thought of, e.g., the current pandemic):

- We are aware that the time for taking action is limited.
- We are aware that we need solutions, quick solutions that are more or less optimal.
- We are willing to take risks and make compromises and reduce our requirements for reliability and efficiency.
- We do not trust skeptics who are without solutions but are always against the solutions of others.

Here is just one current example of a low-level construal (people thinking more concretely, the example of Covid-19). We are aware that we need to find a solution as soon as possible, and the only currently logical solution is vaccination. Therefore, scientists have joined forces and developed a vaccine in less than a year. The process of vaccine development and adoption, which took several years in the past, took less than a year. A solution was developed that is more or less optimal, reliable, and risky, despite the warnings of various groups, which we do not trust. A vast majority of people decided to get vaccinated, which is completely understandable from an expert's point of view, given the state of chaos in which we are now.

However, if we are not in a state of chaos, the other principle applies: the more distant an object is from the individual, the more abstractly it will be thought of. In such cases of high-level construal, people think abstractly and in the long term. Awareness is low. We deny the passing of time, we are not willing to make compromises, we doubt the proposed solutions or even deny the existence of a problem (e.g., environmental issues), and the inevitable question arises, whom we can trust. The basis of trust is the realization that various "nonscientists and pseudoscientists" are exploiting science for their own purposes, thereby destroying the trust in science. The gap between scientists and pseudoscientists, influencers, and politicians is clear here. The solutions to this problem are also long term and need to be developed systematically and in the long run, especially by:

- Developing critical thinking and critical decision-making in the whole society (or at least a large part of it, as true believers can never be convinced by the power of arguments).
- Distinguishing facts (logical argumentation, good science) from pseudo-facts (bad science), the problem of interpretation (system 1 or system 2).
- A strategic separation of truth from potential lies (unverified or poorly verified facts) or lies (deception and manipulation) that are created using pseudoscientific methods and through misleading use of various statistical explanations and shortcuts.

Let us try to consider this issue in more detail from a historical point of view.

4 A Historical Understanding of How Science Works

Understanding the process of science can protect people against misinformation. One of the most intriguing stages in the process of science is noticing when something is "weird." The year 2020 has been a historic year, unfortunately, for the most part, not in a good way. There was a historic level of disregard of scientific advice with respect to the Covid-19 virus. But while the events of 2020 may feel unprecedented, the social pattern of rejecting scientific evidence did not suddenly appear this year. There was never any good scientific reason for rejecting the expert advice on Covid, just as there has never been any good scientific reason for doubting that humans evolved, that vaccines save lives, and that greenhouse gases are driving disruptive climate change. To understand the social pattern of rejecting scientific findings and expert advice, we need to look beyond science to history, which tells us that many of the various forms of the rejection of expert evidence and the promotion of disinformation have roots in the history of tobacco (Oreskes, 2020).

Throughout the first half of the twentieth century, most people saw science as something that makes our lives better. Science had deepened our understanding of the natural world, which helped us to cure diseases, light our homes, and bring new forms of entertainment into our lives. Research and development really did produce products that measurably improved many lives. But science was also developing the playbook for science denial and disinformation. The chief culprit in this darker story was the tobacco industry. It's disparaged science by promoting the idea that the link between tobacco use and lung cancer and other diseases was uncertain or incomplete and that the attempt to regulate it was a threat to humans' freedom. The industry made products more addictive by increasing their nicotine content, while publicly denying that nicotine was addictive. With these tactics, the industry was able to delay effective measures to discourage smoking long after the scientific evidence of its harms was clear. The same arguments were used to delay action on acid rain, the ozone hole, and climate change, and in the year 2020, we saw the spurious "freedom" argument being used, for example, to disparage mask wearing or vaccinating. We also saw the "tobacco strategy" seeping into social media, which influences public opinion and which many people feel needs to be subject to greater scrutiny. Climate scientists have complained that the social media giant contributes to the spread of climate denial by permitting false or misleading claims, while hobbling responses by mainstream scientists by labeling their posts "political."

Without a historical perspective, we might interpret this as a novel problem created by a novel technology. Like the tobacco industry, social media companies sold us a toxic product, while insisting that it was simply giving consumers what they wanted.

With climate-related anxiety, stress, and post-traumatic stress disorder on the rise, a contingent of mental health professionals are developing a new standard of mental healthcare for our climate-changed world. Their profession faces a steep learning curve. There is growing recognition in the field of psychology that people are experiencing distress over climate change. More than 40% of Americans felt "disgusted" or "helpless" about climate change, according to a survey published by researchers at Yale University (Leiserowitz et al., 2021).

It is only natural to feel anxious in the face of a melting planet and the sixth mass extinction, both wrought by human actions. But while humanity may be responsible for the carbon pollution warming our planet, the reality is that just a few large corporations – and complicit politicians – have set us on this path. As individuals, it is easy to feel helpless to stop the destruction of the biosphere. Climate change seemed like the ultimate external problem. If I had no control over climate change, how could I begin to tackle my own despair? Is this true? "Eco-anxiety is a natural response to a threat. And this is a very real threat." Climate change is the reality we all live in now. Between 2009 and 2020, the proportion of Americans who said they had personally experienced the effects of global warming increased from 32% to 42%, according to the 2020 survey from the Yale Program on Climate Change Communication. And in some cases, these effects are directly impacting mental health (Leiserowitz et al., 2021).

While the world has a choice when it comes to limiting climate change, magically stopping all carbon pollution tomorrow would still leave decades of warming baked into the system. That means, presumably, the mental health impacts could worsen into the future. Society will have to adapt to many changes, including how we treat the attendant grief and anxiety of life on a less stable planet. "People are existing and dealing with personal problems in the context of this global crisis now. And the global crisis will impact on the way you deal with personal problems."

5 What Can (Must) We Do?

Every vote counts!
Do all votes count the same?

Certainly, the voice of the everyman does not count as much as the voice of an influencer. Let us try to analyze this briefly on data obtained from Forbes magazine analyses. The starting point is the everyman:

1 vote of the everyman counts 1 vote.
1 vote of the most influential politician (10 politicians) counts approx. 790 million votes.
1 vote of one of the world's most powerful people (75) counts 105 million votes.
1 vote of a billionaire (2000) counts approx. 2 million votes.

What can the everyman do?

- Operate on a partial local level (saving energy, living sustainably, separating and reducing waste, living according to the principle of a circular economy).
- Connect and put pressure on decision-makers! (In order to have the impact equal to the impact of one of the most powerful people in the world, the everyman must hypothetically connect with 105 million like-minded people.)

And what can an influencer do?

- Be aware of the problems (obtain and internalize information from the everyman).
- Influence others at their level (if their vote is worth 105 million votes, getting just one like-minded person results in 210 million votes, etc.).
- Negative example: one of the ten most influential politicians in the United States makes a decision on behalf of the 311 million people living in the United States, and the United States abandons all global environmental agreements.
- Positive example: another one out of the ten most influential politicians in the United States made a different decision, also on behalf of the 311 million people living in the United States, and the United States re-accedes to all global environmental agreements.
- Invest in the ideas he or she advocates:
  - Bill Gates has invested 3.5 trillion USD in the development of next-generation batteries.
  - Elon Musk has invested between 6 and 7.5 trillion USD in the development of high-speed rail travel, etc.

We should probably opt for a bottom-up model to address such changes, regardless of how illusory it may appear. We need to be aware that the interests of influencers and the interests of the everyman are usually worlds apart. Influential people are aware that they can "buy" their own world, their own part of the "paradise," where they will create "ideal conditions" at the expense of others. But the everyman will have to live in the (part of the) world that will remain. Our strength, the power of the everyman, is in the masses. But our problem is that we lack the awareness, that we lack faith, that we fail to realize that the only way to achieve change is to join forces, and that with this in mind, we must forget all that divides us and instead connect regardless of differences between us. We should realize that we are really running out of time and that in the optimistic version we are at a deception point, from which there is no return, though pessimists would say that we have already crossed this point a long time ago. We have entered a period of chaos, a period of possible great change. The only problem with this is that we need to really know what we want and be aware that we do not have much time left to take action!

We have two enemies:

1. The *inner one*, ourselves, we are (unfortunately) our greatest enemy.
2. The *outer enemy* – time, which will keep breathing down our necks and forcing us to make decisions. Just think of how we addressed (and are still addressing) the pandemic, and still, the results are not how we would like them to be! The problem, however, is that a permanent solution is possible only in the long term, i.e., creating awareness starting with young people throughout the whole vertical of education.

Let us now try to make sense of the environmental problem in a broader social context.

6 Conclusion

Environmental costs are *costs associated with the actual or potential deterioration of natural assets due to economic activities*, which are reflected in the ecological footprint, i.e., in the human activities measured in terms of the area of biologically productive land and water required to produce the goods consumed and to assimilate the wastes generated. Environmental issues should (must) be observed through CLT. According to CLT, psychological distance affects the extent to which we think about an event, person, or idea:

- *High-level construal is when people think abstractly.*
- *Low-level construal is when people think more concretely and is associated with psychological proximity.*

The solutions to ecological problem (high level construal) advocated in this book are also long term and need to be developed systematically and in the long run, especially by:

- Developing critical thinking and critical decision-making in the entire society (or at least a large part of it, as true believers can never be convinced by the power of arguments).
- Distinguishing facts (logical argumentation, good science) from pseudo-facts (bad science), the problem of interpretation (system 1 or system 2).
- A strategic separation of truth from potential lies (unverified or poorly verified facts) or lies (deception and manipulation) that are created using pseudoscientific methods and through misleading use of various statistical explanations and shortcuts.

Chapter 4
Social Relations

Abstract This chapter, *Social Relations*, addresses the development of social competences and critical thinking and points out that in today's technologically developed world, we need to develop our critical thinking more and more quickly (today's information is worthless tomorrow). This is related to two basic paradoxes, namely:

1. The problems addressed are increasingly more complex and more interdisciplinary, while our actions are increasingly more monodisciplinary.
2. Providing solutions to such problems requires time, but in today's quickly developing world, time is what we don't have.

At any given moment in today's world, we are confronted with imperfect information, information that requires a huge amount of specific expertise, or misleading information (whether unintentionally or intentionally). In the world of false prophets generating misleading information, there are countless believers who blindly believe everything these false prophets have to say, as indeed, true believers would. Based on a social-cognitive perspective, this chapter tackles a wide range of topics including misleading information, attitudes, person perception and person management, social awareness and social management as a part of emotional intelligence, prejudice, stereotypes, self-concept, discrimination, persuasion, decision-making, and other areas.

It can be argued that teaching is something unique, unpredictable, and closely related to a person as an individual in a society. Formal constants can be established for this individual, on the basis of which they are objectivized and formalized. The starting point – or an example of this kind of naturalist approach to the development of human awareness – is the modern philosophy of mind. Society is defined by the interrelations between its elements, i.e., the individuals that constitute it. These interrelations are highly complex and, thus, cannot be addressed in their entirety, which is why this social reality can never be understood entirely. In order to be able to understand society, at least partially, we need to examine how it is influenced by the physical environment, culture, and interpersonal relations, since each of these generates social values and institutions that in return change society; for example, education affects the attitude toward one's surroundings (it ultimately also affects the

B. Aberšek, A. Flogie, *Human Awareness, Energy and Environmental Attitudes*,
https://doi.org/10.1007/978-3-031-05871-4_4

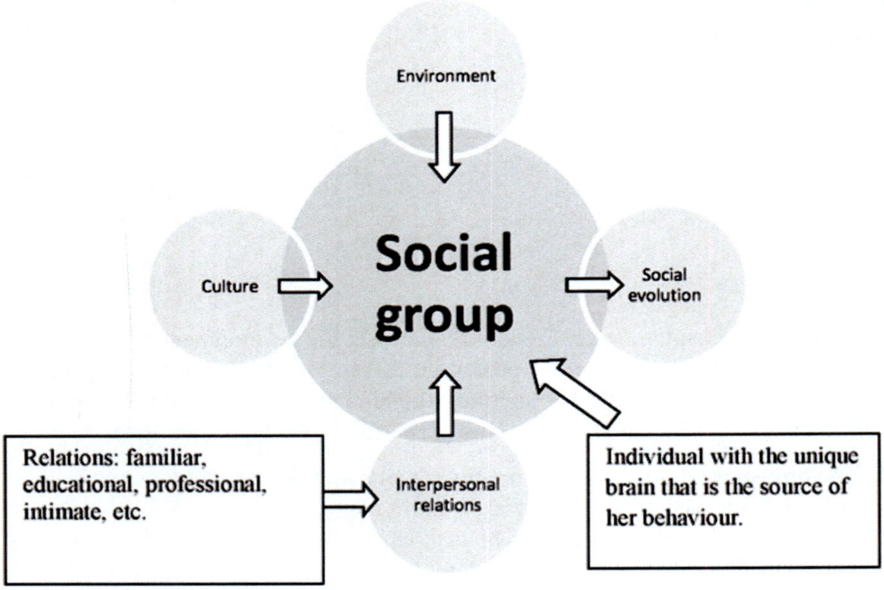

Fig. 4.1 Society as a social system (Aberšek et al., 2014b)

economy) and thereby changes human awareness, cultural relations, and the entire society. In this context, we are mostly concerned with the social development of an individual and their behavior in the specific cause-and-effect relationship between a teacher and student, as is shown in the process of education.

Figure 4.1 schematically shows the cause-consequence effects, with a focus on how the social development of the individual has an impact on the internal and external responses of the individual in society.

In order to balance our behavior, we, human beings, have developed a sophisticated nervous system that informs us of the following:

- The needs of our *internal environment.*
- About what is going on in our *external environment.*

Some of our behavior types are very elementary and do not need to be adapted, since adaptations already happened in the past and they are now simply an automated way of responding to internal and external stimuli. Other, more sophisticated types of behavior demand the recall of pleasant or unpleasant past experiences and the development of a suitable reaction based upon them. These represent most of one's obtained social and cultural knowledge. The third type of behavior demands more elaborate planning, which also includes imagination and an abstract manner of thinking, whereby a strategy that ensures a less unpleasant or painful action is developed (Kahneman, 2011).

We are quite capable of simulating and describing the external environment, using this or that kind of symbol system. However, this becomes much more difficult, or even impossible, when we wish to do the same for our *internal environment*, related to the individual's *consciousness*. Chalmers (1996) divided this problem into two parts:

- *The easy problem* – how do our conscious acts cause the activation of neurons that, in return, cause our behavior?
- *The hard problem*[1] – are these acts at all conscious?

Let us not deal with the hard problem (and leave it to philosophers) and, instead, focus on the easy problem and explore what can be done from the viewpoint of the goal of our book. For this purpose, we will imagine each individual as a *regulated system*, in the sense that the social group, in which these individuals operate, creates a servomechanism, i.e., an external control, which defines different values (prescribes social norms) to the behavior of individuals within this social group (Müller, 2008, 2011). This understanding of an education system means that it is a cybernetic system, which can be formalized (Aberšek et al., 2014b); we will reinforce this assumption with our own arguments later in this book.

1 From the Sociological to the Anthropological Level

To understand the ultimate origin of familiar behaviors, such as walking, aggression, and making love, we must go back to the fundamental laws of physics and specifically the *second law of thermodynamics*. This law states that entropy always increases: a cathedral will end up as a pile of stones, but a pile of stones will never become a cathedral. Consequently, for an organism to stay alive in an environment that is less organized than the organism itself, that organism must constantly seek the energy to combat entropy. It follows that, to paraphrase Henri Laborit, the only reason for being is to be, or according to Maslow's hierarchy of needs (Fig. 4.2), surviving is the elementary human need – in other words, to maintain the body's highly organized structure in an environment where everything tends toward disorder. However, as soon as we are less threatened than we were historically (when our genetic code was generated), the elementary human needs become equally important as our need for comfort and a high quality of life.

Let us now focus only on human safety needs. This includes personal security, employment, resources, health, and prosperity. From the point of view of CLT, safety needs can be divided into two groups:

- Low-level construal is when people think more concretely and in the short run. It is associated with psychological proximity, i.e., *personal security* and

[1] Also called *the explanatory gap*.

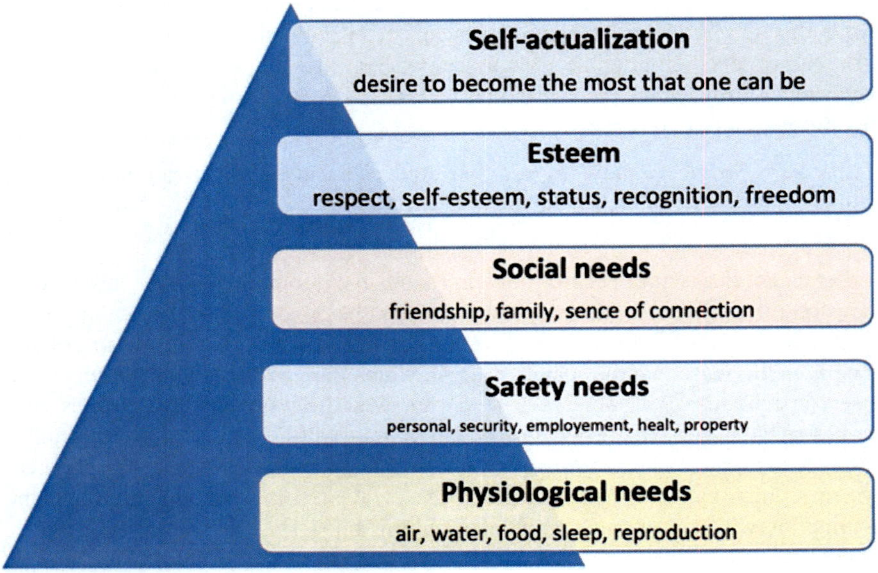

Fig. 4.2 Maslow's hierarchy of needs

employment (potentially also *resources* and *health*, if we are old and vulnerable (e.g., the Covid-19 crises and illnesses)).
- High-level construal is when people think abstractly and in the long term, for example, thinking about *health* (if we are young and healthy) and especially *prosperity*.

Most often we set short-term, near, and concrete goals, because we know that the proximity and concreteness motivate us (we know that we can achieve this goal). We do not burden ourselves too much with distant goals, especially if they are abstract and require us giving up some of the easier, more concrete, and closer goals (e.g., we want a beautiful and comfortable life today, and we do not think about the impact of our actions on the future, e.g., the impact on our environment).

As mentioned earlier, living organisms have developed two major ways of obtaining energy:

1. Drawing it directly from the sun's rays, this is what *plants* do when they use *photosynthesis* to produce energy from solar entropy;
2. Evolving *behaviors* to secure solar energy that plants have already converted into edible form, this is what *animals* do, including human beings.

To guide these behaviors, animals have developed a nervous system that:

3. Informs them of the needs of their *internal environment*.
4. Let's them know what is happening in their *external environment*.

An animal perceives this external world through its senses. The animal's brain then processes this sensory information to construct a certain representation of this world, drawing on memory to attribute pleasant or not-so-pleasant affective values to it. After that comes the behavior as such. It involves a movement whereby the animal acts on its environment, so as to maintain its integrity as an organism.

The human, or his nervous system, throughout the history of all time will be adjusted to current needs and possibilities. Perhaps the human is the most adaptable creature, as it is able to work and live in all of them, even the most impossible environments. This environment is also able to adapt and in this process is its most important property, the *ability to learn, the ability to develop different awareness*.

Education, and the method of schooling and studying, does not provide the youth with a competitive advantage, i.e., employability, anymore. Therefore, today, the young must be equipped with fundamental competences and awareness that are general in character (e.g., learning how to learn) and are thus transferable between different areas (generic competences). Skills, and technical-vocational knowledge, must be added to this (UNESCO, 2002). The results of the *Progress in International Reading Literacy Study* (PIRLS, 2016) suggest that the model of quality teaching is composed of three fundamental *dimensions*, each of them being composed of six *elements*:

- *Intellectual dimension* (elements, deep knowledge, deep understanding, problematic knowledge, higher-order thinking, metalanguage, and substantive communication).
- *Learning environment* (elements, explicit quality criteria, commitment, high expectations, mutual support, students' self-control, and student-teacher joint decisions).
- *Making learning meaningful* (elements, prior knowledge, cultural sophistication, knowledge integration, inclusion, and narration).

In analyzing and comparing different international studies, it is important to consider the fact that differences in student performance, across OECD member countries, are explained with the following factors (UNESCO, 2008):

- 55% by student characteristics
- 34% by environmental factor influence
- 11% by differences between countries.

1.1 Human as a Social Being

Human beings are social beings, which means that their social competences are equally if not more important than their cognitive skills. When speaking about social competences, we are talking also about emotional intelligence (EI) and emotional competences. There are many different definitions and explanations of the term competence (OECD, 2005). The notion becomes even more complex when speaking particularly of social competences. The object of this explanation is not to analyze the entire area of social competences; therefore, one of the most general definitions of the term is used: social competences refer to the social, emotional, and cognitive skills and behaviors that children need for successful social adaptation (Semrud-Clikeman, 2007), noting that social competence is also defined as the ability to handle social interactions effectively. Our interest will focus mainly on self-awareness and social awareness as part of emotional intelligence (see Fig. 4.10). Social competence and/or social skills are used here as general terms defining the complex social interactions in society. One of the fundamental questions, to which this book tries to provide answers, is: How do people build awareness, and how can society promote certain types of awareness? Surely, one of the most logical answers is by means of education! In the following section, we discuss the process of cognitive and affective development of an individual. One of the basic theories in this field is Bloom's taxonomy (Bloom et al., 1956).

To be a good teacher you need to acquire a large set of different kinds of interdisciplinary knowledges and skills, each of which binds an idea of the causal relation, *cause and consequences*, possible antecedents and causes, possible developments and consequences, and possible interventions to strategy of teaching.

The causes and consequences (effects, impacts), for example, of historical events (such as the industrial revolution), are of primary interest to researchers, and this relationship is an important part of various kinds of research. Therefore, it is important to think about effective ways of teaching and understanding these concepts. Figuring out why a historical or any other event happened can be complicated, as there are often a range of different causes. Similarly, there are always multiple consequences of an event. One way of understanding causes and consequences is through categorizing them. For example, here are three categories which researchers (and students engaged in research-based learning) can use to support their understanding (Table 4.1).[2]

A deeper understanding of judgments and choices also requires a richer vocabulary than is available in everyday language. Learning is closely connected with the body-mind relation, with our way of thinking. When you are asked what you are thinking about, the answer is usually not very complicated. You believe you know what goes on in your mind, which often consists of one conscious thought leading in an orderly way to another. But that is not the only way the mind works, nor indeed is that the typical way. Most impressions and thoughts arise in your conscious experience without your knowing how they got there. You cannot imagine how you came to the belief that everyone loves you, or how you detected a hint of irritation in your spouse's voice on the telephone, or how you managed to avoid a threat on the road before you became consciously aware of it. The mental work that produces impres-

[2] https://nzhistory.govt.nz/te-akomanga/historical-concepts/cause-consequence-analysis

Table 4.1 Causes and consequences through categorizing

Chronology	Nature/ type	Significance/values
Long term	Ideological	This can be analyzed using 6 Rs criteria[a]:
Short term	Political	Responsibility, resourcefulness, resilience, respect, reflection, reasoning
Underlying (specific to causes)	Social	
Catalyst/trigger (specific to causes)	Economic	
Ongoing (specific to consequences)	Positive	
	Negative	

[a]As a school we educate and provide boundaries for our students to make choices safely, through of provision of a safe environment with the support of the strong foundations formed from our core values (*responsibility*, *resourcefulness*, *resilience*, *respect*, *reflection*, and *reasoning*)

sions, intuitions, and many decisions goes on in silence in our mind. It is like the mind has its own thought and language, the language of thought.

Much of the discussion in this chapter will relate to our thought, our judgments, and our brain – especially in relation to the learning process and to awareness. Most of our judgments and actions are appropriate most of the time. As we navigate our lives, we normally allow ourselves to be guided by impressions and feelings, and the confidence we have in our intuitive beliefs and preferences is usually justified. But not always. We are often confident even when we are wrong, and an objective observer is more likely to detect our errors than we are. This chapter mostly presents our current curiosity, judgment, and decision-making, which have been shaped by cognitive and neuroscientific discoveries of recent decades.

Historians of science have often noted that at any given time scholars in a particular field tend to share basic assumptions about their subject. Social scientists are no exception; they rely on a view of human nature that provides the background of most discussions of specific behaviors but is rarely questioned. Social scientists in the 1970s broadly accepted two ideas about human nature. *First*, people are generally rational, and their thinking is normally sound. *Second*, emotions such as fear, affection, and hatred explain most of the occasions on which people depart from rationality. We documented systematic errors in the thinking of normal people, and we traced these errors to the design of the machinery of cognition rather than to the corruption of thought by emotion. Scholars in other disciplines found it useful, and the ideas of heuristics and biases have been used productively in many fields, including intelligence analysis, philosophy, statistics, education, and teaching/learning strategy. A recurrent theme is that *proprioception* plays a large role in every successful learning story; it is almost always easy to identify a small change in the story that would have turned a remarkable achievement into a mediocre outcome (Aberšek, 2018).

1.2 Where We Are Now

Our main aim here is to present a view of how the mind works that draws on recent developments in cognitive and social psychology. One of the more important developments is that we now understand the marvels as well as the flaws of intuitive thought (Aberšek, 2016; Kahneman, 2011).

When confronted with an ecology and environmental problem, the machinery of intuitive thought does the best it can. If the individual has relevant expertise, he/she will recognize the situation, and the intuitive solution that comes to his/her mind is likely to be correct. This is what happens when an environment protection expert looks at a complex position: the few ideas that immediately occur to him are all strong. When the question is difficult and a skilled solution is not available, intuition still has a shot: an answer may come to mind quickly – but it is not an answer to the original question. The question faced (should I invest in nuclear sector stock?) was difficult, but the answer to an easier and related question (do I have a positive attitude to nuclear energy?) came readily to mind and determined the choice. This is the essence of intuitive heuristics: when faced with a difficult question, we often answer an easier one instead, usually without noticing the substitution.

The spontaneous search for an intuitive solution sometimes fails: neither an expert solution nor a heuristic answer comes to mind. In such cases we often find ourselves switching to a slower, more deliberate, and effortful form of thinking. Fast thinking includes both variants of intuitive thought – the expert and the heuristic – as well as the entirely automatic mental activities of perception and memory, the operations that enable you to know, for example, "if everyone likes you," or retrieve the name of the capital of Slovenia.

The distinction between simple and complex thinking has been explored by many psychologists over the last 25 years. Kahneman (2011) described mental life by using a metaphor of two agents, called *system 1* and *system 2*, which, respectively, produce fast (simple) and slow (complex) thinking. Kahneman speaks of the features of intuitive and deliberate thought as if they were traits and dispositions of two characters in the human mind. In the picture that emerges from recent research, the intuitive system 1 is more influential than your experience tells you, and it is the secret author of many of the choices and judgments you make. Most of this book is about the workings of system 1 and the mutual influences between it and system 2.

System 1 continually constructs a coherent interpretation of what is going on in our world at any instant. We attempt to give a sense of the complexity and richness of the automatic and often unconscious processes that underlie intuitive thinking and of how these automatic processes explain the heuristics of judgment. The goal is to introduce a language for thinking and talking about the mind. Why is it so difficult for us to think in a complex manner, statistically? We easily think associatively, we think metaphorically, we think causally, but statistics requires thinking about many things at once, which is complex and complicated, something that system 1 is not designed to do. The difficulties of statistical thinking describe a puzzling limitation of our mind: our excessive confidence in what we believe we know and our apparent inability to acknowledge the full extent of our ignorance and the uncertainty of the world we live in. We are prone to overestimate how much we understand about the world and to underestimate the role of chance in events. Overconfidence is fed by the illusory certainty of hindsight (Kahneman, 2011).

There is a distinction between *two selves*, the experiencing self and the remembering self, which do not have the same interests. For example, we can expose people to two painful experiences. One of these experiences is strictly worse than the other, because it is longer. But the automatic formation of memories – a feature of system 1 – has its rules, which we can exploit so that the worse episode leaves a better memory. When people later choose which experience to repeat, they are, naturally, guided by their remembering self and expose themselves (their experiencing self) to unnecessary pain. The distinction between two selves is applied to the measurement of well-being, where we find again that what makes the experiencing self happy is not quite the same as what satisfies the remembering self. How two selves within a single body can pursue happiness raises some difficult questions, both for individuals and for societies that view the well-being of the population as a policy objective. And this is, in fact, the central problem in the development of critical thinking and critical judgment and in the construction of an internal value system. These elements can be best affected through an adaptive system of teaching and learning, or in one word, education.

2 Taxonomies and Learning

The term *taxonomy* was first introduced to the field of learning and education by Bloom in his work *Taxonomy of Educational Objectives: Handbook I, The Cognitive Domain* (Bloom et al., 1956). This work is the fruit of the work of several experts. There has been considerable discussion in Bloom's group about the term *taxonomy*. Despite the opposition of some collaborators, however, this term has been extremely well received, especially among curriculum planners, researchers, and practicing teachers at all levels of education (Anderson & Sosniak, 1994). Thus, the term became established throughout the profession. A number of other taxonomies emerged after Bloom's, but they were all more or less mere "fortunate flirts" with Bloom's.

One of the fundamental principles of modern teaching is a modernized and extended understanding of what knowledge is. Today, knowledge is no longer understood only as a collection of content (facts, information, or elaborated explanations) but, in the broadest sense, as the understanding, use, and connections between these contents, as the ability to create a critical improvement based on such content, and also as a repertoire of skills used in the process, e.g., research, communication, and collaborative skills for working with resources; skills needed for working with modern technologies, especially ICT; skills needed to present ideas; etc. Therefore, when assessing knowledge, we should not only be interested in what content and how much of that content students have mastered but also how they understand this content; how they use it in new situations; how they solve problems; how they compare, conclude, and argue; how they use resources of various kinds; in what ways they present their ideas and findings; how critical and creative they are; and last but not least, how they developed a critical way of thinking, values, and an awareness of various social problems during the learning process.

We often see that curricula at the level of general objectives encourage quality, lifelong knowledge, and higher levels of thinking across different taxonomies, while at the level of operational objectives they slip to lower levels of thinking such as familiarization, imitation, and perhaps even partial understanding. Due to the demands of various expert fields (which wish to preserve a wide range of different contents at any cost), there is not much room for maneuver for more in-depth reflections about the division of operational objectives at different taxonomic levels, which would encourage independent thinking and building knowledge and values at higher cognitive levels.

We must always be aware that if we want to promote a truly active approach of students to learning and attitudes in society, as discussed in previous chapters, then it is crucial to design the treatment of learning content in such a way so that students arrive to knowledge through their own activity and through discovery (constructivist approach, cognitive approach).

Two fundamental questions arise here:

- What are the truly fundamental, indispensable types of knowledge and values that learners will need in their work as well as in everyday life in the future?
- Who is in fact competent to select these types of knowledge for the learning population in the context of their applicability 10 or 20 years from now?

To these, we can, of course, add an endless number of additional dilemmas and/or questions, such as the following:

- Is it possible at all to find such knowledge, ethical values, and awareness, especially with respect to emerging and rapidly evolving technologies and the society as a whole (e.g., Society 5.0 or the "supersmart society")?
- What types of knowledge are these?
- Is it really that important to be teaching mainly content (something which becomes obsolete in a few years)?
- In what way, then, should we design a modern society?
- What does contemporary teaching refer to in our context?
- Etc., etc.

Indeed, most of these questions refer to a systemic, conceptual level; they are aimed at the school system and the agents within it. However, each individual teacher/parent must decide individually what they can accept (and above all, what they want to accept) and what is realistic, acceptable, and useful for themselves and their students/children/learners. Everyone will have to consider the extent to which they feel they are limited by the system in testing and applying new ideas and strategies and the extent to which they are using this as a welcome excuse to not swallow the bitter pill of changing established patterns of thinking and acting. After all, the hardest thing to change is someone's mind!

2.1 *Bloom's Taxonomy*

Bloom's taxonomy (Bloom et al., 1956) is a classification of learning objectives within education, proposed in 1956 by a committee of educators, chaired by Benjamin Bloom, who also edited the first volume of the standard text, *Taxonomy of Educational Objectives: The Classification of Educational Goals* (referred to as simply, "the Handbook"). Although named after Bloom, the publication followed a series of conferences from 1949 to 1953, which were designed to improve communication between educators on the design of curricula and examinations.

It refers to a classification of the different objectives that educators set for students (learning objectives). Bloom's taxonomy divides educational objectives into three "domains":

- *The cognitive domain* – the domain of knowledge, which includes six classification levels.
- *The affective domain* – the domain of attitudes and emotions consisting of five classification levels.
- *The psychomotor domain* – the domain of skills, which includes six classification levels.

This concept is schematically shown in Fig. 4.3.

Let us now consider a somewhat simplified schema; see Fig. 4.4.

Within the domains, learning at the higher levels is dependent on having attained prerequisite knowledge and skills at lower levels. A goal of Bloom's taxonomy is to motivate educators to focus on all three domains, creating a more *holistic* form of education.

In short, these three areas could be summarized as: skills-knowledge-attitudes or work-think-feel. This structure is logical and quite old. Aristotle already defined a similar structure of a good speaker, which is characterized by three appeals:

Learning domain	Categories					
Cognitive	Knowledge	Comprehension	Application	Analysis	Synthesis	Evaluation
Affective	Acceptance	Response	Values	Organization or conceptual values	Internalization or characterization of values	
Psychomotor	Imitation	Manipulation	Precision	Articulation (inference)	Naturalization (adaptation)	

Fig. 4.3 Bloom's taxonomy domains

Cognitive knowledge	Affective attitudes	Psychomotor skills
1. Recalling information	1. Awareness	1. Imitating, duplicating
2. Understanding	2. Response, reaction	2. Manipulation (following instructions)
3. Applying	3. Values (understanding and acting)	3. Development of precision
4. Analyzing (structures/elements)	4. Organizing a personal value system	4. Articulation (combining, integrating appropriate skills)
5. Synthesis (build, create)	5. Internal value system (behavior)	5. Naturalization (routine, becoming an expert)
6. Evaluation (assessment, judgment on a topic)		

Fig. 4.4 Simplified Bloom's taxonomy

- *Logos*, the cognitive area, logical way of arguing, the head.
- *Pathos*, the affective area, emotions, the heart.
- *Ethos*, which stands out somewhat from the above scheme, but only seemingly. We could say that the charisma of a speaker is closely related to his skills/abilities, i.e., to their psychomotor domain.

Bloom's research dedicated to the cognitive domain was completed in 1956 (Bloom et al., 1956). For the other two domains, however, he left some research space for other authors as well. Anderson, Krathwol, Masla, Simpson, Harrow, and Dave developed other taxonomies, which is a sign of high-quality research. The last three in particular focused on the psychomotor domain. Let us now focus only briefly on the cognitive domain, which has received the most attention and is also the basis for understanding the other two taxonomies.

2.2 What Is Bloom's Taxonomy?

Bloom's taxonomy is a multilevel model classifying the learning process into six levels depending on the complexity of the process. Over years of development, these levels have often been portrayed as steps that encourage both teachers and students to climb to the very top of the thinking process. The three lowest levels are *knowledge*, *comprehension*, and *application*. The three highest levels are *analysis*, *synthesis*, and *evaluation*. The taxonomy is built hierarchically, which means that if we master the higher level, we also master all the lower levels. For example, if we know how to use something, it means that we also understand how it works and have the necessary knowledge. From this we can easily detect the natural division into lower and higher cognitive levels.

Bloom's taxonomy in all its various forms represents, to put it simply, the learning process. In some cases, it is simplified, but it still clearly shows this hierarchical structure of learning. To summarize:

- Before we can *understand* a concept, we have to *remember it.*
- Before we can *apply* a concept, we need to *understand it.*
- Before we can *analyze* a concept, we must be able to *apply it.*
- Before we can *evaluate* its effects, we must *analyze them.*
- Before we can *create*, we must *remember, understand, analyze*, and *evaluate.*

Of course, not every learning process necessarily starts with memorization and does not necessarily contain all the categories and all the phases of this process. It is also not necessary to always reach the highest level of thinking.

Bloom's taxonomy, especially for the cognitive and affective domains, has withstood all the upheavals of history and has survived to this day only in a slightly modified form. The basic idea and basic starting points remained the same.

2.3 Revised Bloom's Taxonomy

In the 1990s, a former Bloom student, Lorin Anderson, adapted the original taxonomy for use in the twenty-first century. At the same time, Anderson put together a multidisciplinary team made up of cognitive psychologists, curriculum theorists, researchers, and testing and assessment specialists (Anderson et al., 2001). Like the core group, this one too took 6 years to complete its work, which was published in 2001. This revised taxonomy, while retaining all its original starting points, nevertheless brings several important innovations. For those who would like to study the starting points and the resulting changes in detail, there are several excellent sources. Here we focus only on summarizing of the most important ones. The changes could be divided into three groups:

1. Terminology changes.
2. Structural changes.
3. Changes in emphasis.

2.4 Terminology Changes

Changes in terminology between the two versions are perhaps the most obvious differences and can also cause the most confusion. Basically, Bloom's six major categories were changed from noun to verb forms. Additionally, the lowest level of the original, knowledge, was renamed and became remembering. Finally, comprehension and synthesis were retitled to understanding and creating. To minimize the confusion, compare the two images shown in Fig. 4.5.

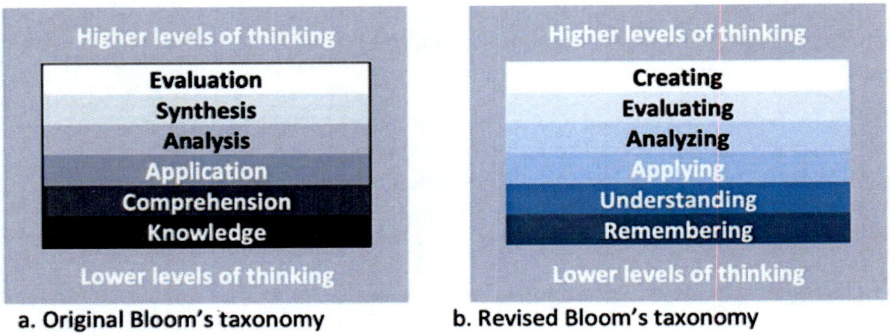

Fig. 4.5 Genesis of Bloom's taxonomy

The new terms in the revised Bloom's taxonomy are defined as follows (Anderson et al., 2001):

Remembering: Retrieving, recognizing, and recalling relevant knowledge from long-term memory

Understanding: Constructing meaning from oral, written, and graphic messages through interpreting, exemplifying, classifying, summarizing, inferring, comparing, and explaining

Applying: Carrying out or using a procedure through executing or implementing

Analyzing: Breaking material into constituent parts, determining how the parts relate to one another and to an overall structure or purpose through differentiating, organizing, and attributing

Evaluating: Making judgments based on criteria and standards through checking and critiquing

Creating: Putting elements together to form a coherent or functional whole and reorganizing elements into a new pattern or structure through generating, planning, or producing

2.5 *Structural Changes*

Structural changes seem dramatic at the first sight, but upon a closer look, they are quite logical. Bloom's original cognitive taxonomy was a one-dimensional form. The revised Bloom's taxonomy takes the form of a two-dimensional table. One of the dimensions identifies the knowledge dimension (or the kind of knowledge to be learned), while the second identifies the cognitive process dimension (or the process

Knowledge dimension	Cognitive process dimension					
	Remember	Understand	Apply	Analyze	Evaluate	Create
Factual knowledge	List	Summarize	Classify	Order	Rank	Combine
Conceptual knowledge	Describe	Interpret	Experiment	Explain	Assess	Plan
Procedural knowledge	Tabulate	Predict	Calculate	Differentiate	Conclude	Compose
Meta-cognitive knowledge	Appropriate use	Execute	Construct	Achieve	Action	Actualize

Fig. 4.6 Bloom's revised taxonomy table

used to learn). As represented in Fig. 4.6, the intersection of the knowledge and cognitive process categories forms 24 separate cells.

The knowledge dimension on the left side is composed of four levels that are defined as factual, conceptual, procedural, and meta-cognitive. Each of the four knowledge dimension levels is subdivided into either three or four categories (e.g., factual is divided into factual, knowledge of terminology, and knowledge of specific details and elements).

Perhaps the only "arguable" point seems to be the substitution of categories 5 and 6, i.e., synthesis (create) and evaluation. Expert opinions on this are divided. If we consider the classification globally, at a strategic level, then the highest-order skill should contain a strategic evaluation (effective, management, strategic assessment, and decision-making), whereby such strategic evaluation logically belongs to the category of evaluation.

2.6 *Changes in Emphasis*

Emphasis is the third and final category of changes. As noted earlier, Bloom himself recognized that the taxonomy was being "unexpectedly" used by countless groups never considered an audience for the original publication. The revised version of the taxonomy is intended for a much broader audience. Emphasis is placed upon its use as an authentic tool for curriculum planning, instructional delivery, and assessment.

2.7 *Why Use Bloom's Taxonomy?*

Bloom's taxonomy provided educators with one of the first systematic classifications of the processes of thinking and learning. The cumulative hierarchical framework requires achievement of the prior skill or ability before the next, more complex one and remains easy to understand. In simple terms, you cannot understand a concept unless you remember it first. Similarly, you cannot use knowledge or a concept unless you understand it, etc. The structure thus connects lower-order levels with higher-order levels. Bloom's taxonomy remains a very effective tool for measuring cognitive abilities.

With the dramatic changes in society over the past few decades, the revised Bloom's taxonomy provides an even more powerful tool to fit today's teachers' needs. The structure of the revised taxonomy table matrix "provides a clear, concise visual representation" (Anderson et al., 2001) of the alignment between standards and educational goals, objectives, products, and activities. Today's teachers must make tough decisions about how to spend their classroom time. Clear alignment of educational objectives with local, state, and national standards is a necessity. Like pieces of a huge puzzle, everything must fit properly. Bloom's taxonomy makes this possible.

2.8 *How to Use Bloom's Taxonomy?*

A search of the Internet yields clear evidence that Bloom's taxonomy has been applied to a variety of situations, including the field of multiple intelligences, problem-solving skills, creative and critical thinking, and recently even technological integration. Because of the various possible ways of using the taxonomy, it may be difficult to provide uniform instructions on how to use it, but since there are a number of good practice examples in literature, one is able to find the appropriate form.

2.9 *Bloom's Taxonomy for Affective Domain*

A lot has been said about the cognitive domain taxonomy already; let us now turn our attention to the fundamental domain that is relevant to our interpretations, i.e., the affective domain according to Bloom at Fig. 4.7.

This domain was researched in detail by Bloom, Krathwhol, and Masia in 1964, who published *Taxonomy of Educational Objectives: Volume II, The Affective Domain* (Bloom et al., 1956). As with the other domains, the affective domain detail provides a framework for teaching, training, assessing, and evaluating the effectiveness of training and lesson design and delivery.

This domain may be a little trickier to understand than the others. The differences between the levels, especially between 3, 4, and 5, are subtle and not so clear as the separations elsewhere in the taxonomy. Therefore, it is very useful to observe this part of the taxonomy holistically and not partially, and above all, affective taxonomy is closely related to the philosophy of emotional intelligence. In this way, we arrive at a comprehensive theory for understanding people in certain unstable conditions, such as the attitude toward the environment, ecological awareness, global warming, environmental pollution, etc. Each of these areas requires a systematic approach,

Level	Category or 'level'	Behavior descriptions	Examples of experience, or demonstration and evidence to be measured	'Key words' (verbs which describe the activity to be trained or measured at each level)
1	**Receive**	Open to experience, willing to hear	Listen to teacher or trainer, take interest in session or learning experience, take notes, turn up, make time for learning experience, participate passively	Ask, listen, focus, attend, take part, discuss, acknowledge, hear, be open to, retain, follow, concentrate, read, do, feel, perceive
2	**Respond**	React and participate actively	Participate actively in group discussion, active participation in activity, interest in outcomes, enthusiasm for action, question and probe ideas, suggest interpretation	React, respond, seek clarification, interpret, clarify, provide other references and examples, contribute, question, present, cite, become animated or excited, help team, write, perform
3	**Value**	Attach values and express personal opinions	Decide worth and relevance of ideas, experiences; accept or commit to particular stance or action	Argue, challenge, debate, refute, confront, justify, persuade, criticize
4	**Organize or conceptualize values**	Reconcile internal conflicts; develop value system	Qualify and quantify personal views, state personal position and reasons, state beliefs	Build, develop, formulate, defend, modify, relate, prioritize, reconcile, contrast, arrange, compare
5	**Internalize or characterize values**	Adopt belief system	Self-reliant; behave consistently with personal value set	Act, display, influence, solve, practice

Fig. 4.7 Bloom's affective domain taxonomy

cause-and-effect analyses, a systematic structural and functional analysis of cause-and-effect relationships, and, of course, critical thinking, critical judgment and attitude, or awareness. However, since these relations are not clear-cut and unambiguous (they are not black and white), it is necessary to introduce systems of optimization and cost-benefit analysis in such systems.

Most of what we have learned before regarding the cognitive domain also applies here, namely:

- Bloom's affective taxonomy in all its various forms represents, in simple terms, the process of acquiring and consolidating certain values and awareness. In some cases, this taxonomy is simplified, but it still vividly illustrates the hierarchical structure of learning and, for the affective domain, the acquisition and consolidation of certain values and awareness.
- We cannot reach a higher-order thinking level unless we have previously mastered a lower-order thinking level, i.e., if we do not acknowledge a certain situation, for example, an environmental issue, then we will not (cannot) respond to it properly.
- The taxonomy of the affective domain is also built hierarchically, which means that if we master a higher-order thinking level, we have also mastered all lower-

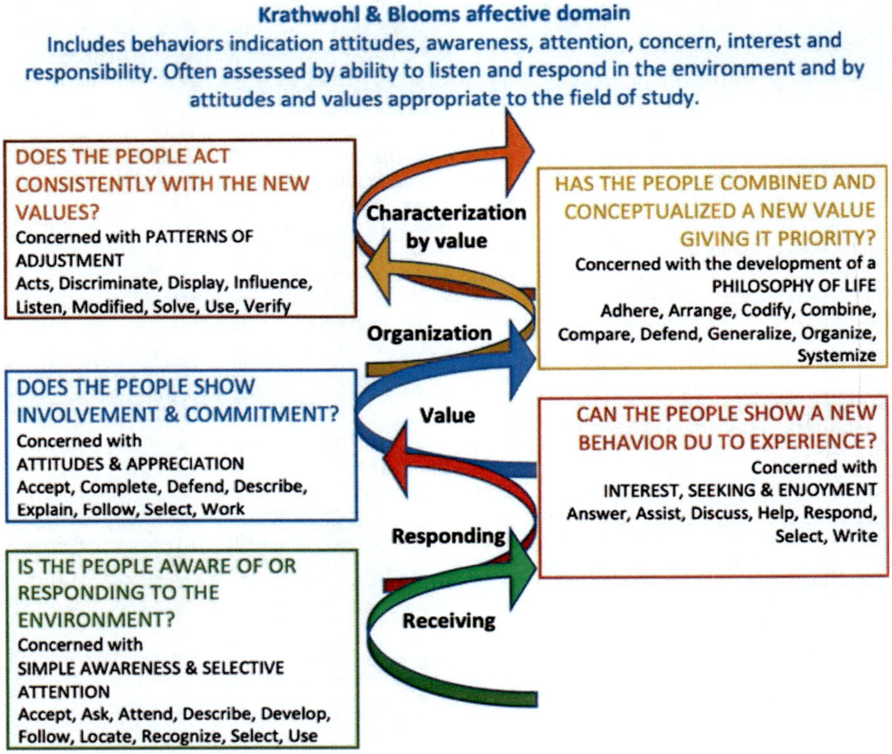

Fig. 4.8 Bloom's affective domain taxonomy

order levels (we respond in accordance with our internalized values). For instance, if we have created values (environmental protection, environmental awareness), this means that we respond to situations related to these values, we are aware of environmental problems and their consequences, and we are ready to accept them as our own (and we feel ready to give up something for the sake of resolving this issue in the future).

Bloom's taxonomy for the affective domain is symbolically shown in Fig. 4.8.

3 Emotional Intelligence

As mentioned earlier, one notion that could be seen as part of what Bloom called the affective domain is *emotional intelligence*. Let us consider this in light of environmental issues.

The themes discussed in this book include energy production, the ecological footprint, global warming, and ecology, in relation to self-awareness and social

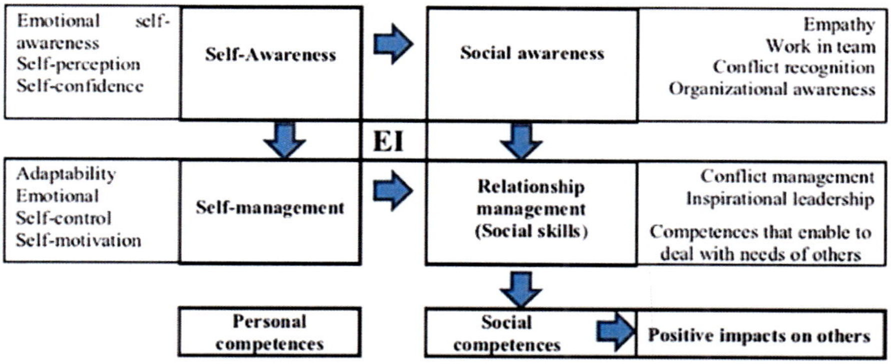

Fig. 4.9 Emotional intelligence (EI) and social competence

awareness. With social awareness and social skills, the emotional intelligence of a person (Fig. 4.9) also develops, changing, step by step, the human historical memory and adding new elements to this historical memory on the level of intuitive thinking (system 1 according to Kahneman). Social awareness should be developed in every individual, first, as part of their primary social environment (family) and afterward as part of their secondary environment (school). A specific way of thinking (creative, critical, and conscious thinking) should be encouraged in the development of every person, and it is very important to begin this process with learners (children) of the youngest possible age. With regard to the issue of *ecological footprint*, social but also cognitive competence should be developed step by step, helping to raise awareness about the fact that we have only one Earth, one planet, which is like a complex puzzle, for which everyone is responsible and to which everyone can contribute their small (but important) piece.

In Fig. 4.9, *self-awareness* refers to knowing one's internal states, perception, resources, and intuition; *self-management* refers to managing one's internal states, impulses, and resources (Bloom's affective domain); *social awareness* refers to how people handle relationships and to the awareness of others' feelings, needs, and concerns, while *relationship management* is concerned with the skill or adeptness at inducing desirable responses in others (Aberšek, 2018; Goleman, 2007).

To be aware of one's own thought is the way to critical thinking and a higher taxonomical level of knowledge. In problem- and research-based learning, people automatically develop critical thinking processes. Moreover, apart from critical thinking processes, emotional intelligence, social skills, and competences are also developed when working and conducting research in groups, with the aim of solving a problem together, as a team (Bohm, 2004; Kahneman, 2011).

A central dilemma of human life is the dilemma between self and social awareness/management. Obviously, this dilemma troubles us more than ever during the age of environmental degradation problems. Philosophers call this *the problem of other minds*. We could also refer to it as the *solipsism*[3] *problem*. Solipsism, technically, is an extreme form of skepticism, at once utterly strange and irrefutable. It holds that you are the only conscious being in existence. The cosmos sprang into existence when you became sentient, and it will vanish when you die. As crazy as this proposition seems, it rests on a brute fact: each of us is sealed in an impermeable prison cell of subjective awareness. In reality, our experience, our own mind, exists only in our head, while we can only infer the existence of other minds indirectly. The Chinese room experiment[4] (Searle, 1980) serves as a metaphor not only for the world of physics (or computer sciences) but also for the human condition. Each of us sits alone within the cell of our subjective awareness. Now and then we receive cryptic messages from the outside world. Only dimly comprehending what we are doing, we compose responses, which we slip under the door. In this way, we manage to survive, even though we *never really know what the hell is happening*. We find ourselves in such situations more often today than ever before.

However, before we embark on addressing the problem of other minds (solipsism), let us focus on the philosophical explanation of why it is so hard to change people's minds. Let us connect all the theories discussed so far at an anthropological level, from Bloom's affective domain taxonomy, through emotional intelligence, the construal level theory and Maslow's hierarchy of needs, to philosophical skepticism and a perspective on the problem of other minds (solipsism). Let us consider all these at the level of the relation(s) between personal and social awareness (see Fig. 4.10).

4 Solipsism and the Problem of Other Minds

If solipsism speaks only of the "self" (anthropological perspective), then we could say that emotional intelligence speaks of the "self in relation to others" (sociological perspective). Now, let us turn our attention to the problem at hand, i.e., solipsism and the problem of other minds. Solipsism is sometimes expressed as the view that "I am the only mind which exists" or "my mental states are the only mental states." Solipsism is therefore more properly regarded as the doctrine in which "existence" means for me *my* existence and that of *my* mental states. Existence is everything that I experience – physical objects, other people, events and processes, and my relation to nature. Existence is anything that would commonly be regarded as a constituent of

[3] *Solipsism* (from Latin *solus* "alone" and *ipse* "self") is the philosophical idea that only one's mind is sure to exist. As an epistemological position, solipsism holds that knowledge of anything outside one's own mind is unsure; the external world and other minds cannot be known and might not exist outside the mind.

[4] The *Chinese room* argument holds that a digital computer executing a program cannot have a "mind," "understanding," or "consciousness," regardless of how intelligently or humanlike the program may make the computer behave. The point of Searle's experiment is to make a non-Chinese man simulate a native Chinese speaker in such a way that there wouldn't be any distinction between these two individuals.

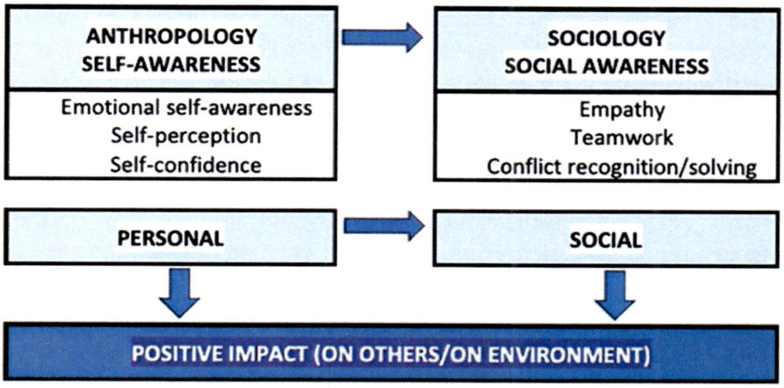

Fig. 4.10 Relation between personal and social awareness

the space and time in which I coexist with others and is necessarily construed by me as part of the content of *my consciousness*. For the solipsist, it is not merely the case that he believes that his thoughts, experiences, and emotions are, as a matter of contingent fact, the only thoughts, experiences, and emotions. Rather, the solipsist can attach no meaning to the supposition that there could be thoughts, experiences, and emotions other than his own. In short, the true solipsist understands the word "pain," for example, to mean "my pain," the word "nature" to mean "my nature (the natural environment in which I live)." The solipsist cannot accordingly conceive how this word is to be applied in any sense other than this exclusively egocentric one. To apply this to the problem at hand, the solipsist understands, for example, the destruction of something (a part of the environment) that is "his own" or perhaps at best the destruction of a part of the environment close to him (e.g., a thunderstorm or a flood). However, it is much harder for him to understand – or he does not, in fact, understand at all – distant, global problems (e.g., glaciers melting).

As a theory, if indeed it can be termed as such, it is clearly very far from common sense. Therefore, it might reasonably be asked why the problem of solipsism should receive any philosophical attention or why we mention it in the context of environmental issues and environmental awareness. There are two answers to this question.

Firstly, while no great philosopher has explicitly espoused solipsism, this can be attributed to the inconsistency of much philosophical reasoning. Many philosophers have failed to accept the logical consequences of their own most fundamental commitments and preconceptions. The foundations of solipsism lie at the heart of the view that the individual gets his own psychological concepts (thinking, willing, perceiving, aware, and so forth) from "his own case," that is, by abstraction from "inner experience." In this sense, solipsism is implicit in many philosophies of knowledge and mind since Descartes, and any theory of knowledge that adopts the Cartesian egocentric approach as its basic frame of reference is inherently solipsistic.

Secondly, solipsism merits close examination because it is based upon three widely entertained philosophical presuppositions, which are themselves of fundamental and wide-ranging importance. These are as follows:

(a) What I know most certainly are the contents of my own mind – my thoughts, experiences, affective states, and so on.
(b) There is no conceptual or logically necessary link between the mental and the physical. For example, there is no necessary link between the occurrence of certain conscious experiences or mental states and the "possession" and behavioral dispositions of a body of a particular kind.
(c) The experiences of a given person are necessarily private to that person, from a constructive philosophical point of view.

These presuppositions are of undeniable Cartesian origin and are widely accepted by philosophers and nonphilosophers alike. In addressing the problem of solipsism, one immediately grapples with fundamental issues in the philosophy of mind. However spurious the problem of solipsism per se may strike one, these latter issues are unquestionably important. Indeed, one of the merits of the entire enterprise is the extent that it reveals a direct connection between apparently unexceptionable and certainly widely held common sense beliefs and the acceptance of solipsistic conclusions. If this connection exists and we wish to avoid those solipsistic conclusions, we shall have no option but to revise, or at least to critically review, the beliefs from which they derive logical sustenance.

There is a temptation to say that solipsism is a false philosophical theory, but this is not quite strong or accurate enough. As a theory, it is only incoherent. From a practical standpoint, however, i.e., from the experience of its opponents "I don't agree as such," it describes and explains very well the attitude of these individuals toward various problems, such as "impending problems" (the pandemic of Covid-19, the attitude toward vaccination), or toward different environmental issues. After all, the human race in its essence is a selfish and self-centered race, which is willing to destroy a lot for its own brief moment of glory, for its own "small benefit." Other people *seem* to possess conscious perceptions, emotions, memories, and intentions, just as we do, but we can't be sure they do. We can only guess how the world looks to others, based on our behavior and utterances, including these words we are reading, but we have no firsthand access to others' inner life. For all we know, it might be a mindless bot.

Natural selection instilled in us the capacity for a so-called theory of mind – a talent for intuiting others' emotions and intentions. But we have a countertendency to deceive each other and to fear we are being deceived. The ultimate deception would be pretending we are conscious when we are not. The solipsism problem thwarts efforts to explain consciousness. Scientists and philosophers have proposed countless contradictory hypotheses about what consciousness is and how it arises. Panpsychists contend that all creatures and even inanimate matter – even a single proton – possess consciousness. Hard-core materialists insist, conversely, that not even humans are all that conscious. The solipsism problem prevents us from verifying or falsifying these and other claims. We cannot be certain that we are conscious, let alone a jellyfish, robot, or chatbot as Siri or Alexis. As long as we lack what neuroscientist Christof Koch calls a *consciousness meter*,[5] a device that can measure consciousness in the same way that a thermometer measures temperature, theories of consciousness will remain in the realm of pure speculation.

But the solipsism problem is far more than a technical philosophical matter. It could be a paranoid but understandable response to the feelings of solitude that lurk within us all. Even if you reject solipsism as an intellectual position, we can sense it, emotionally, whenever we feel estranged from others and whenever we confront the awful truth that we can never know, *really know* another person, and no one can really know us.

Religion is one response to the solipsism problem. Our ancestors dreamed up a supernatural entity who bears witness to our innermost fears and desires. No matter how lonesome we feel, how alienated from our fellow humans, someone is always there watching over us. Of course, there are many different religions and beliefs out there, from Christianity and Buddhism to modern ones like Scientology. And to deny environmental problems, we must be just as deeply religious as those who believe that there exists a supernatural force, some supernatural being who created us and everything around us. We are running away from responsibility, away from being the creator of our own environment; we are returning to a mythology, to the prescientific period of interpretation. We are not truly responsible for anything; someone else is to blame for our misery and for our failure; someone else will solve our (environmental) problems, whether they are scientists, politicians, or just anyone (e.g., a divine creator). However, even if we are not believers, we can still deny environmental issues (e.g., like Trump). We do not see the problem, and we do not apprehend it unless we actually face that problem immediately (e.g., hurricanes, weather disasters, and other nature's calls) from the standpoint of our rationality. Yet we all need to believe in something, and we all make decisions based on incomplete knowledge and uncertainty, which may turn out to be – right or wrong.

5 Decision-Making, Incomplete Knowledge, and Uncertainty

Psychologists study how humans make decisions by presenting them with problems in the form of brain games about, for example, energy and the environment: cheap energy or green (environmentally friendly) energy – which would you rather have? Most people would choose the first option. Brain teasers are one thing, but we live in

[5] Consciousness is everything you experience. The origin and nature of these experiences, sometimes referred to as qualia, have been a mystery from the earliest days of antiquity right up to the present. Many modern analytic philosophers of mind, most prominently perhaps Daniel Dennett of Tufts University, find the existence of consciousness such an intolerable affront to what they believe should be a meaningless universe of matter and the void that they declare it to be an illusion. That is, they either deny that qualia exist or argue that they can never be meaningfully studied by science.

a real world with real problems, not make-believe ones. People are sometimes faced with situations that require them to make life-and-death decisions in the face of incomplete or uncertain knowledge. Instead of trying to change people's minds, we set about learning how they *really* think about these problems. To that end, we can ask them questions designed to reveal how they assess risks. The answers help us understand why people form beliefs about divisive issues such as nuclear energy – and today, climate change – when they do not have all the facts. We want to figure out how well the general public understands the risks they face in everyday life. And yet, the explanation is relatively simple, if only we remember Kahneman's system 1 and system 2. We know that humans in their essence are lazy beings!

5.1 Understanding Risk

Did experts always have an accurate understanding of the probabilities for disaster? Experts analyze risks by breaking complex problems into more knowable parts. When scientific understanding is incomplete, risk analysis shifts from reliance on established facts to expert judgment. Studies of those judgments find that they are often quite good but only when experts get good feedback. Until scientists don't get good feedback, they will be unsure and will not know – how accurate their risk estimates really are (Drummond & Fischhoff, 2017).

5.2 Thinking of the Unthinkable

One of the more important developments in cognitive and social psychology is that we now understand the beauty as well as the flaws of intuitive thought, known according to Kahneman as system 1. When confronted with a problem – e.g., choosing to buy an electric car or deciding whether to invest in a stock – the machinery of intuitive thought does the best it can. If the individual has relevant expertise, they will recognize the situation, and the intuitive solution that comes to their mind is likely to be correct. This is what happens when an expert looks at a complex position: the few moves that immediately occur to him are all strong. When the question is difficult and a skilled solution is not available, intuition still has a shot: an answer may come to mind quickly – but it is not an answer to the original question. The question faced (should I behave environmentally friendly?) was difficult, but the answer to an easier and related question (do I like the car?) came readily to their mind and determined their choice. This is the essence of intuitive heuristics: when faced with a difficult question, we often answer an easier one instead, usually without noticing the substitution (Kahneman, 2011).

The spontaneous search for an intuitive solution sometimes fails – neither an expert solution nor a heuristic answer comes to mind. Now, consider here two ways of thinking: one is slow, automatic, we refer to it as system 1, and one is fast, system

2. There is a distinction between the automatic operations of system 1 and the controlled operations of system 2. Associative memory, the core of system 1, continually constructs a coherent interpretation of what is going on in our world at any instant. These complex automatic and often unconscious processes underlie intuitive thinking and, in this way, explain the heuristics of judgment.

In different cases we often find ourselves switching to a slower, more deliberate, and effortful form of thinking. Fast thinking includes both variants of intuitive thought – the expert and the heuristic – as well as the entirely automatic mental activities of perception and memory, the operations that enable you to know if everyone likes you or retrieve information about what the ecological footprint is. The distinction between fast and slow thinking has been explored by many psychologists over the last few decades. The features of intuitive and deliberate thought are to be thought of as if they were traits and dispositions of two characters in one's mind. In the picture emerging from recent research, the intuitive system 1 is more influential than personal experience, and it is the key to many of the choices and judgments we make (Kahneman, 2011).

How, then, can researchers in this area fulfill their duty to inform the public about accurate ways to think about events and choices that are beyond their experience? Scientists can, in fact, accomplish this if they follow two basic lessons from studies of decision-making.

5.3 *Lesson 1*

The facts of climate science will not speak for themselves. The science needs to be translated into terms that are relevant to people's decisions about their lives, their communities, and their society (system 2). While most scientists are experienced communicators in a classroom, out in the world, they may not get feedback on how clear or relevant their messages are. Addressing this feedback problem is straightforward: *test messages before sending them*. One can learn a lot simply by asking people to read and paraphrase a message. When communication researchers have asked for such rephrasing about weather forecasts, for example, they have found that some are confused by the statement that there is a "70 percent chance of rain." The problem is with the words, not the number. Does the forecast mean it will rain 70 percent of the time? Over 70 percent of the area? Or there is a 70 percent chance of at least approx. 1 mm of rain at the weather station? The last interpretation is the correct answer.

Sometimes people decide intuitively (system 1) but many times want to know more than the probability of rain or ecological footprint when they make decisions. They want to understand the processes that lead to those probabilities: how things work (system 2). Studies have found that some critical aspects of climate change research are not intuitive for many people (system 1), such as how scientists can argue yet still agree about the threat of climate change or how carbon dioxide is different from other pollutants. (It stays in the atmosphere longer!) People may reject the research results unless scientists tell them more about how they were derived. Interpretation and communication should be as close to people as possible, e.g., inform them about natural disasters in their vicinity and not some far away case from central China or India (even if natural disasters are of more catastrophic dimensions in that part of the world). The arguments should take into account the psychological distance and be more concrete than abstract, as we have learned from the section on construal level theory.

5.4 Lesson 2

People who agree on the facts can still disagree on what to do about them. A solution that seems reasonable to some can seem too expensive or unfair to others. For example, people who like plans for carbon capture and sequestration, because it keeps carbon dioxide out of the air, might oppose using it on coal-fired power plants. They fear an indirect consequence: cleaner coal may make mountaintop-removal mining more acceptable. Those who know what cap-and-trade schemes are meant to do – create incentives for reducing emissions – might still believe that they will benefit banks more than the environment. These examples show why two-way communication is so important in such situations. We need to learn what is on others' minds and make them feel like partners in decision-making. Sometimes that communication will reveal misunderstandings that research can reduce. Or it may reveal solutions that make more people content.

5.5 More than Science

How facts are communicated and interpreted are important because climate-related decisions are not always based on what research says or shows. For some individuals, scientific evidence or economic impacts are less important than what certain decisions reveal about their beliefs. These people ask how their choice will affect the way others think about them, as well as how they think about themselves.

For instance, there are people who forgot energy conservation measures but not because they are against conservation. They just do not want to be perceived as eco-freaks. Others who conserve do it more as a symbolic gesture and not based on a belief that it makes a real difference. Knowing the science does not necessarily mean agreeing with the science.

5.6 *When Decisions Matter Most*

Although decision science researchers still have much to learn, their overall message about ways to deal with uncertain, high-stakes situations is optimistic. When scientists communicate poorly, it often indicates that they have fallen prey to a natural human tendency to exaggerate how well others understand them. When laypeople make mistakes, it often reflects their reliance on mental models that have served them well in other situations but that are not accurate in current circumstances. When people disagree about what decisions to make, it is often because they have different goals rather than different facts.

In each case, the research points to ways to help people better understand one another and themselves. Communication studies[6] can help scientists create clearer messages. And decision science[7] can help the public to refine their mental models to interpret new phenomena. By reducing miscommunication and focusing on legitimate disagreements, decision researchers can help society have fewer conflicts and make dealing with the ones that remain easier for us all.

6 Conclusion

Society is defined by the interrelations between its elements, i.e., the individuals that constitute it. These interrelations are highly complex and, thus, cannot be addressed in their entirety. In order to be able to understand society, at least partially, we need to examine how it is influenced by the physical environment, culture, and interpersonal relations, since each of these generates social values and institutions that in return change society; for example, education affects the attitude toward one's surroundings and thereby changes human awareness, cultural relations, and the entire society. In this context, we are mostly concerned with the social development of an individual and their behavior in the specific cause-and-effect relationship between a teacher and student, as is shown in the process of education.

Throughout the course of human history, the only reason for human existence has been existence (i.e., being) itself, or, in the jargon of Maslow's hierarchy of needs,

---

[6] *Communication studies or communication science* is a discipline that deals with processes of human communication and behavior, patterns of communication in interpersonal relationships, social interactions, and communication in different cultures. Communication is commonly defined as giving, receiving, or exchanging ideas, information, signals, or messages through appropriate media, enabling individuals or groups to persuade, to seek information, to give information, or to express emotions effectively.

[7] *Decision science* is the collection of quantitative techniques used to inform decision-making at the individual and population levels. It includes decision analysis, risk analysis, cost-benefit and cost-effectiveness analysis, constrained optimization, simulation modeling, and behavioral decision theory, as well as parts of operations research, microeconomics, statistical inference, management control, cognitive and social psychology, and computer science.

surviving is an elementary human need. However, as soon as we are less threatened than we were historically, the elementary human needs become equally important as our need for comfort and a high quality of life. In this situation, from the point of view of CLT, safety needs can be divided into two groups:

- Needs at the level of low-level construal, i.e., when people think more concretely and in the short run. This kind of thinking is associated with psychological proximity, i.e., *personal security*, *employment*, and *health* (e.g., the Covid-19 crisis and illnesses).
- Needs at the level of high-level construal, i.e., when people think abstractly and in the long term, for example, thinking about *prosperity* or *ecology*.

This chapter presented recent developments in cognitive and social psychology. When confronted with an environmental problem, the machinery of intuitive thought does the best it can. If the individual has relevant expertise, they will recognize the situation, and the intuitive solution that comes to their mind is likely to be correct. When the question is difficult and a skilled solution is not available, intuition still has a shot: an answer may come to mind quickly – but it is not an answer to the original question. The question faced (should I invest in nuclear sector stock?) was difficult, but the answer to an easier and related question (do I have a positive attitude to nuclear energy?) came readily to mind and determined the choice. This is the essence of intuitive heuristics: when faced with a difficult question, we often answer an easier one instead, usually without noticing the substitution.

Developing social competence and critical thinking is of key importance.

In today's technologically developed world, we need to develop our (critical) thinking more and more quickly (information is here today and trash tomorrow). At every step, we are confronted with imperfect information, information that requires a huge amount of specific expertise, or misleading information (whether unintentionally and intentionally). In the world of false prophets generating misleading information, there are countless believers who blindly believe everything these false prophets have to say.

Social cognition is not simply a topic within social psychology; it is an approach to studying any subject with social psychology. Using a social-cognitive perspective, researchers can study a wide range of topics including attitudes, person perception and person management, social awareness and social management as a part of emotional intelligence, prejudice, stereotypes, self-concept, discrimination, persuasion, decision-making, and other areas.

Chapter 5
Energies and Energetics

Abstract This chapter, *Energies and Energetics*, deals with the fundamental knowledge and basic definitions from the fields of primary forms of energy, energy transformation, and energy production. We point out that energy is the capacity to do work and that it occurs in nature in various forms. Energy can neither be created from nothing, nor it can be destroyed (used up, exhausted). It can only be transferred or changed from one form to another. Energetics is the study of energy, the origin of energy, and the flow and transformation of energy. Further on, the chapter discusses topics such as electricity mix, energy consumption, and sustainable developement.

Energy is the capacity to do work. It occurs in nature in various forms. Energy can neither be created from nothing, nor it can be destroyed (used up, exhausted). Energy can only be transferred or changed from one form to another. *Energetics* is the study of energy, the origin of energy, and the flow and transformation of energy.

When we talk about energy management, we often use phrases such as "efficient use of energy," "rational use of energy," etc., although it would be more correct to refer to it as "usage" or "exploitation" of energy. Energy is a quantity that cannot be "used up," which is a fact described by the *law of conservation of energy*, which states that:

> Energy can neither be created nor destroyed – only converted from one form of energy to another.

The energy of a body is its capacity to do work. Human beings do not have a sensory organ for directly perceiving energy, but we can detect the effects of energy – for example, the movement of bodies, radiated heat, light, sound, and similar things. Energy can be measured through appropriate devices and instruments. The unit for measuring the quantity of energy is joule, and we also use its derivatives (kJ, MJ, PJ, etc.). In everyday life we are perhaps more familiar with another possible unit for measuring energy, the watt-second (Ws, $1\text{J} = 1\text{Ws}$) and its derivatives Wh, kWh, and MWh. The old units, which are not allowed in the International System of Units, include calorie, kilocalorie, erg, kilopond-meter, and others.

B. Aberšek, A. Flogie, *Human Awareness, Energy and Environmental Attitudes*,
https://doi.org/10.1007/978-3-031-05871-4_5

We can "use up" thermal energy in the sense that we may convert it to a different form or level but we cannot use up (destroy) energy in general. This is why we use the term "use" of energy, despite the slight inconsistency regarding the exact meaning of that term. Some languages distinguish between different adjectives to describe or refer to things or concepts directly related to energy.

1 Basic Forms of Energy

All forms of energy can be classified into two basic groups:

- Accumulated energy in a physical space or body
- Transient energy, which occurs when accumulated energy passes from one form to another or when potential energy passes from one body to another

Accumulated energy is durable, i.e., it can be preserved in a specific form for any length of time, while transitions are the basic characteristic of *transition energy*.

2 Accumulated Energy

Accumulated energy comes in three main forms:

- Potential energy
- Kinetic energy
- Internal energy

Potential Energy To lift an object of mass m (in kilograms) through an initial height h_0 (in meters) to a final height h_1 (in meters), it is necessary to do the work W_p (in joules):

$$W_p = m\ g(h_1 - h_2) \tag{5.1}$$

whereby g is the acceleration of gravity. Its value is $g = +9.81$ m/s^2. By lifting this mass, the work done is not lost, as it can be retrieved if we allow the body to fall again to the height of h_0 due to its weight. Work accumulates in the lifted object as potential energy, or stored energy. This energy is actually a consequence of the mutual attraction between the Earth's mass and the mass of the lifted body and depends largely on their relative position.

Kinetic Energy In order to accelerate a body of mass m from velocity v_0 to velocity v_1, it is necessary to do the work:

$$W_k = \frac{1}{2}m\left(v_1^2 - v_0^2\right) \tag{5.2}$$

This work, however, can be restored if the moving mass returns to the initial velocity v_0. In short, we can assume that the work is accumulated in the object in the form of kinetic energy.

Internal Energy Potential and kinetic energy can accumulate not only in objects as a whole but also in the smallest parts (elements) of these objects. For example, gas molecules are in constant motion. Among them, there are attractive and repulsive forces, indicating the existence of potential energy. As molecules move, this potential energy is converted into kinetic energy, but upon collision between the molecules, it is transformed back into potential energy. Molecules also oscillate in rigid bodies, which results in a similar energy transformation as in gases. The sum of this energy, if there are no external influences, is constant. Likewise, forces act in atoms and in the nucleus of an atom, resulting in the existence of energies. A huge amount of energy is accumulated especially in an atom's dense nucleus.

All these forms of energy, be it energy at the level of a molecule, an atom, or a nucleus, represent *the internal energy of matter*. The internal energy at the *level of the molecule* changes with the supply or dissipation of heat, which causes an increase or reduction in the speed of movement of molecules. This energy is called *internal thermodynamic energy*. Internal energy at the atomic level is called *chemical energy* because it refers to energy stored in the bonds of chemical compounds, which my change (e.g., combustion). We distinguish between chemical changes in which the chemical energy decreases, which is reflected in the increase of the internal thermodynamic energy and chemical changes that can occur only when substances are supplied with a certain amount of energy.

The internal energy *at the level of the atom's nucleus* is called *nuclear energy*. There is a distinction between nuclear fusion energy and nuclear fission energy. In the following sections, we focus on the nuclear energy and its transformations. Although ways of exploiting such energy are still being explored, fusion energy is the energy of the future that could solve all current problems, both in terms of obtaining energy and protecting the environment.

3 Energy Sources

Energy sources, sometimes called energy carriers or energy currencies, can be divided into *renewable* and *nonrenewable* sources. The former is constantly being replenished, while the latter have limited supplies and will sooner or later be depleted by humanity. Renewable sources include solar energy (although it is in fact a nonrenewable source, we consider it as renewable only from the point of view of the transience of human existence) in the form of direct radiation (heat), biomass, wind, water energy, and geothermal energy. Nonrenewable sources are fossil fuels, nuclear energy, and the energy of chemical reactions from mineral sources (Bradford, 2018).

Non-Renewable	ENERGY	Renewable
Chemical Fossil fuels (Combustion) **Nuclear** Uranium (Fission of atoms) Light elements (Fusion)	**Chemical** **Nuclear** **Gravitational** **Electromagnetic** **Mechanical** **Thermal** **Electrical**	**Chemical** Muscular (Oxidization) **Nuclear** Geothermal (Thermal conversion) Fusion (of Hydrogen) **Gravitational** Tidal (Mechanical) **Indirect Solar** Biomass (Photosynthesis) Wind and Hydraulic (Mechanical) **Direct Solar** Photovoltaic cell (Conversation) Solar thermal (Thermal conversation)

Fig. 5.1 Sources and evolution of energy

Most of the "raw" energy sources are not directly applicable in the sense of satisfying the human needs, which means we need to convert them into usable forms (most commonly electricity, heat, and kinetic energy), using various devices such as:

- Heating boilers (the energy in fuels is converted into thermal energy)
- Various kinds of power plants (hydroelectric power plants, thermal power plants, nuclear power plants, geothermal power plants, tidal power plants, wind turbines, small-scale hydropower plants, etc.)
- Steam engines (energy in fuels is converted into kinetic energy)
- Internal combustion engines (energy in fuels is converted into kinetic energy)
- Electric generators (kinetic energy is converted into electricity)
- Solar receivers (electromagnetic radiation from the sun is converted into thermal energy)
- Photovoltaic cells (electromagnetic radiation from the sun is converted into electricity)
- Heat pumps (electricity and energy of the environment at a lower temperature are converted into heat energy at a higher temperature) (Fig. 5.1).

4 Transition Energies

As mentioned earlier, transition energies are of short duration and represent only the transformation of one form of accumulated energy into another. They can be divided into the following:

- Mechanical work and mechanical energy
- Electrical energy
- Heat

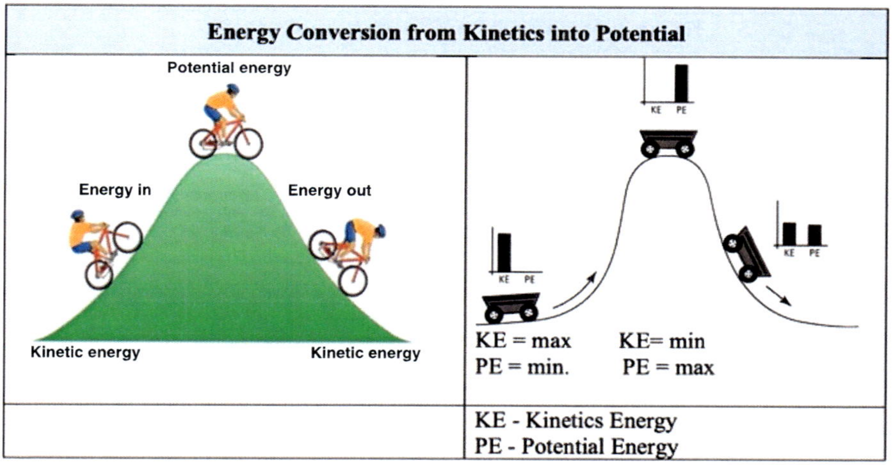

Fig. 5.2 An example of converting kinetic energy into potential energy

Mechanical Work and Mechanical Energy These forms of energy occur, for example, when converting kinetic energy (driving up hill) into the potential energy Fig. 5.2.

Thus, kinetic energy is first converted into mechanical energy, which is used to driving up hill, and is thus gradually converted into potential energy. Therefore, mechanical energy is only a transition energy that cannot accumulate but is transformed further as soon as it occurs.

Electrical Energy We can imagine that bicycle is connected to an electric motor (EM), which receives electricity from the battery. In EM, electrical energy is converted into mechanical energy, which drives the bike up hill.

Heat is also a transition energy, because it is transformed, for example:

- From the combustion gases onto a medium in a steam boiler
- From combustion gases to air in the room when heated by fuel combustion in room furnaces
- From a suitable medium to the coolant, e.g., in the steam turbine condenser, etc.

In all cases, the internal thermodynamic energy of the substance from which the heat comes decreases, and the internal thermodynamic energy of the substance onto which the heat passes increases. The diagram in Fig. 5.3 showing the energy transformation can serve as an illustration.

In this process, heat appears twice as a transition energy: first, in the transfer of heat from the combustion gas onto water and steam (in boiler) and, secondly, in the transfer of heat through the expansion of steam, e.g., in a steam turbine, onto the cooling water in a steam condenser.

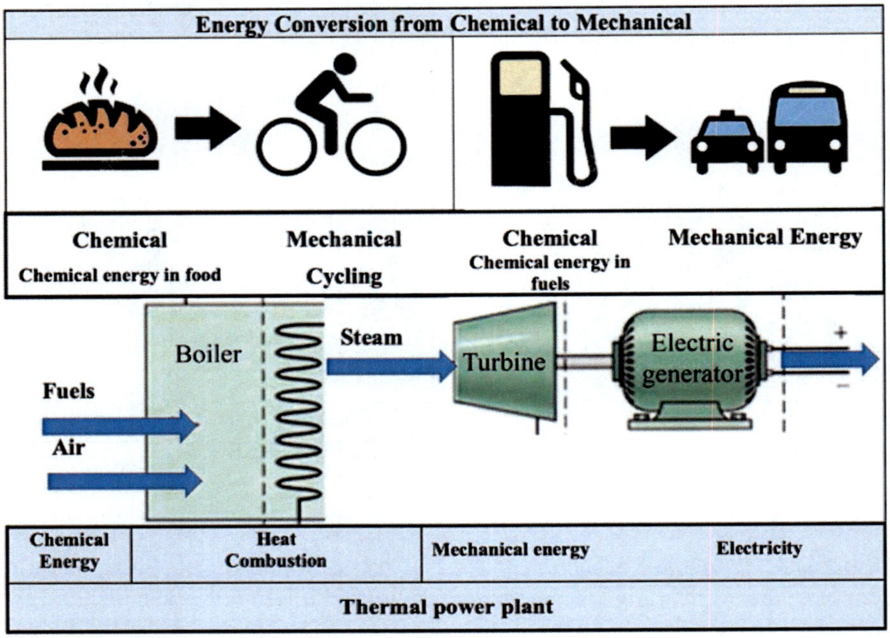

Fig. 5.3 Scheme of energy transformation from chemical to mechanical energy

Due to the heat transfer, the internal amount of energy of combustion gases or expanded steam decreases, while the internal thermodynamic energy of the water, the water vapor, and the coolant increases.

5 Power

Power P is defined as the differential equation of energy and time. Power is a scalar physical quantity defined as the work done in a unit of time. In mechanical systems in which the change in energy is equal to the received or transmitted part, the power is equal to the derivative of the work W after time t:

$$P = \frac{dW}{dt} \tag{5.3}$$

In linear motion, the power is equal to the scalar product of the velocity of the force grip and the force component in the direction of the velocity:

$$P = Fv \tag{5.4}$$

In rotation, the power is equal to the scalar product of torque and angular velocity:

$$P = F\omega \tag{5.5}$$

It can be seen from the equation that power corresponds to the speed of energy utilization or to the speed of energy transformation from one form to another.

6 Primary Forms

Primary forms of energy can be further divided according to the following:

1. *Frequency of use.* We distinguish between *conventional* and *unconventional* forms of energy. Conventional primary forms of energy, the most commonly used forms today, include wood, coal, crude oil, natural gas, hydropower (potential energy of rivers), nuclear fission energy (nuclear fuels such as uranium and thorium), and hot springs (heat appearing on the surface).

   Unconventional primary forms of energy include the following:

   - Oils from oil shale and bismuth sand
   - Kinetic energy of the wind
   - Tidal energy (potential energy)
   - The heat of the Earth's core occurring on the surface
   - Solar energy
   - Internal thermodynamic energy of the sea (exploitation of the temperature difference between the surface and the depth of the sea)
   - Wave and tidal energy
   - Fusion energy of light atoms

2. *The physical properties of the carrier.* We distinguish between the following:

   - Chemical energy carriers (wood, coal, crude oil, oil shale and bismuth sand oils, and natural gas)
   - Nuclear energy carriers (nuclear fuels, light atoms used for fusion)
   - Kinetic energy carriers (wind)
   - Carriers of internal thermodynamic energy (thermal springs, internal energy of the sea)
   - Radiation energy carriers (solar energy)

   Of the primary forms of energy, only hot springs (geysers) can be directly exploited, e.g., to heat spaces. In all other cases, the primary form of energy must be converted into a different, more suitable form of energy.

3. *Renewability.* In this respect, primary forms of energy are divided into the following:

   - *Renewable energy sources*, such as solar radiation, water power, wind energy, tidal and wave energy, and ocean heat. These forms of energy cannot be

exhausted over time, but they can be fully utilized (e.g., the construction of multiple hydropower plants that utilize the full potential energy of a water source).
- *Nonrenewable energy sources*, which are stored in, for example, fossil fuels. Despite the huge quantities, such energy sources can become exhausted.

Most renewable energy cannot be stored – accumulated (wind energy, solar energy, etc.) – so it must be used directly. Certain resources can be accumulated, e.g., water in reservoirs (artificial dams), but only in limited quantities. In contrast, nonrenewable energy sources can be used as needed, creating constant power. An electricity system must have both: permanent energy sources that create constant power, complemented with renewable energy sources such as solar energy and the wind.

7 Useful Forms of Energy

To be able to use different forms of energy, we need to convert it into forms that are useful for humans: into thermal (or heat energy), mechanical, light, and chemical energy (Chapman, 1989).

Thermal Energy (Heat) One of the oldest ways of receiving heat is certainly to directly receive the radiant energy of the sun. However, because this source is not constant and is not available especially when it would be most needed (at night or during winter), several other ways of generating heat have developed throughout history. For a very long time, people received heat mostly by burning wood and fossil fuels. Today, heat can be supplied to us in the form of water vapor or hot water, which are carriers of internal thermodynamic energy. We receive heat with the help of heat exchangers (radiators, etc.). This way of heat transfer is temperature limited. We can also receive heat with the help of electrical energy, which is converted into heat in resistance furnaces or induction furnaces.

Mechanical Energy Today, obtaining work through mechanical energy is most often carried out with the help of electric motors, where electrical energy is converted into mechanical energy (into work). A significant part of mechanical energy is used in transport (use of internal combustion engines, steam boilers, steam turbines). Although mechanical energy can be used directly, in most cases, it is harnessed through an intermediate phase of electrical energy (marine diesel-electric propulsion, diesel-electric locomotives, etc.).

Light Energy Today, we use only electrical energy to produce lighting or, in general, the energy required for telecommunication devices (radio, telephone, television).

Chemical Energy It is mainly used in various technological processes of iron extraction, ferroalloys, electrolysis of aluminum, etc. Coke and electrical energy, or both, are usually required to create chemical processes.

8 Limitations in Energy Conversion

With regard to the possibilities of converting different forms of energy, there are three types of energy:

(a) Energy that can be converted into another form of energy in unlimited quantities is called *exergy*. This type includes potential, kinetic, mechanical, and electrical energy. The complete conversion of such energies is possible only through reversible processes.
(b) Energies that can only be partially converted into other forms of energy (exergy) are internal thermodynamic energy and heat. The limitation of the conversion is a consequence of the second main law of thermodynamics and depends on the form as well as on the state in which the conversion is performed, i.e., it depends on the conditions in the environment (pressure and temperature).
(c) Energy that cannot be converted into another form is called *anergy*. This type includes the energy accumulated in the environment and the energy of all things in the state of the environment.

With the introduction of the concepts of exergy and anergy, the law of conservation of energy reads as follows:

In all energy processes, the sum of exergy and anergy remains constant.

Therefore, each energy consists of *exergy E* and *anergy B:*

$$W = E + B = \text{constant} \tag{5.6}$$

With regard to energy conversion, it therefore applies that:

(a) In all irreversible processes, exergy is converted into energy.
(b) Only in reversible processes does exergy remain constant.
(c) It is impossible to convert anergy into exergy.

Since real energy processes are more or less irreversible, energy conversion reduces the supply of exergy. The terms exergy and anergy have great technical significance. All energy processes (heating, cooling, transportation, processing of materials, etc.) require energy, and not just any type of energy, it has to be exergy (energy that can be converted into other forms). Therefore, energy sources would be more properly called *exergy sources*. Terms such as energy consumption, energy loss, etc. conflict with the law of conservation of energy, according to which energy cannot be destroyed (or lost). In contrast, terms like exergy consumption or exergy loss do have a meaning in physics – because exergy is consumed and lost through conversion into irreversible energy.

9 Energies and Energy Sources

The term *energy source* refers to a substance from which useful energy can be obtained directly or by appropriate conversion. Both energies and energy sources can be divided into the following:

- *Primary energy* – is the energy hidden in energy carriers/energy sources (oil, gas, coal, wood, nuclear energy).
- *Secondary energy* – is the energy obtained through conversion from primary (e.g., electrical energy from coal in a thermal power plant). Conversion losses need to be considered.
- *Final energy* – is the energy received by the user. Conversion losses and transfer losses (these occur on the way from the source to the end user).
- *Useful energy* – is energy which meets the needs of the user, e.g., heat on an electric hob. Losses that need to be considered: conversion losses of the producer, transmission losses, and losses occurring in the conversion of electrical energy to heat (directly at the location of the user).

Let us consider, for example, the path of energy from coal through electricity to the electric motor – losses in the power plant are approximately 60% (secondary energy is 40% of the primary energy), losses in transmission are 10% (final energy is 36% of the primary energy), and losses in the electric motor conversion are an additional 10% of the final energy (useful energy is 32% of the primary energy). For every kilowatt hour of useful work at the output of the electric motor, we therefore used 3 kWh of primary energy from coal. The efficiency of the whole process is therefore 0.32, or 32%. We will come back to this when we discuss the latest "fashion" trend, i.e., electrification. Electrification refers to the transition to electrical energy in almost all segments of our daily lives and to the (necessary or unnecessary) replacement of already established energy sources in a particular area, especially in the introduction of electric mobility. More and more is being said about green energy sources, and electricity certainly is one such resource; however, we seem to be forgetting the process as a whole, that is, how we arrived to this resource in the first place and what was (is) the environmental and economic price we had to pay for it.

10 Energy Content of Some Fuels

Different energy carriers contain different amounts of chemical energy (which is measured in MJ/kg). Some examples are shown in Table 5.1.

Different fuels have different levels of energy content, which can be measured in terms of equivalent energy released through combustion. The higher the energy content the higher the fuel quality, which is inversely proportional to its chemical complexity. High-quality fuels are gases, while low-quality fuels are solids, with liquids in between. The highest energy content fuel is hydrogen, which is also the

Table 5.1 Chemical energy content of some fuels (in MJ/kg)

Fuels	Energy contents (MJ/kg)
Lithium battery	0.5
Wood	17.1
Ethanol	19.9
Methanol	31.1
Coal – anthracite	31.4
Crude oil	41.9
Diesel	45.3
Car gasoline	45.8
Kerosene	46.3
Natural gas	47.2
Propane	50.3
LNG	55.0
Methane	55.5
Hydrogen	142.0

simplest chemical component in existence. Gasoline, which is derived from refining crude oil, contains much more energy than coal (twice than the lower grade bituminous) or wood (three times). Liquid natural gas (LNG) is almost entirely composed of methane, while natural gas has about 85% of its mass accounted by methane. Jet A-1 is the standard fuel used by commercial jet planes and is mostly composed of kerosene and several additives (antifreeze, antioxidant, and antistatic) since the fuel must meet very strict specifications as it will be exposed to high altitudes and low temperatures. Conversely, Bunker C fuel, which is the main fuel used for maritime shipping, can be considered as one of the lowest-quality fuels in liquid form but suitable for the vast ship engines.

Although methane and hydrogen both have higher energy content than gasoline, their gaseous form creates storage difficulties. Furthermore, hydrogen must be synthesized, which requires energy. As a comparison, with a conversion rate of 100%, it would require 100 h to capture the solar energy equivalent of 1 kg of gasoline on a surface of 1 m^2. One of the most efficient energy storage devices, the lithium battery, can only hold about the equivalent of 0.5 MJ per kilogram, underlining the challenge of developing electric vehicles.

Thus, the more energy consumed, the greater the amount of work realized, and it comes as no surprise that economic development is correlated with higher levels of energy consumption. There are four types of physical work related to human activities:

- *Modification of the environment*. It is all the activities involved in making space suitable for human activities, like clearing land for agriculture, modifying the hydrography (irrigation), and establishing distribution infrastructures, as well as constructing and conditioning (temperature and light) enclosed structures.
- *Appropriation of resources*. It involves the extraction of agricultural resources from the biomass and raw materials (minerals, oil, lumber, etc.) for human needs. It also includes the disposal of wastes, which are in an advanced industrial society very work-intensive to dispose of safely (e.g., collection, treatment, and disposal).

- *Processing resources*. It concerns the modification of products from the biomass, of raw materials and of goods, to manufacture according to economic needs. Since the industrial revolution, work related to processing resources was considerably mechanized, initially with simple machines, then assembly lines, and currently with automation.
- *Transfer*. Energy is the potential that allows the mobility of passengers and freight from one location to another. It aims to attenuate the spatial inequalities in the location of resources by overcoming distance. The less energy costs per ton or passenger – kilometer – the less transfers are an economic burden. Overcoming space in a global economy requires a substantial amount of energy and has consequently been subject to massive economies of scale.

There are enormous reserves of energy able to meet the future needs of humanity. Unfortunately, one of the leading contemporary issues is that many of these reserves are not necessarily widely available at competitive costs, such as solar energy, or are unevenly distributed around the world, such as oil and wind energy. Still, the availability or the competitiveness of an energy source can improve with technological development. Even if some energy sources are extracted far from where they are consumed, the massification of transportation enables their mobility.

Through the history of energy use, the choice of an energy source depended on several *utility factors* that involved a transition in energy systems from solid, liquid, and eventually to gas sources. Since the industrial revolution, efforts have been made for work to be performed by machines, which considerably improved industrial productivity. The development of the steam engine and the generation and distribution of electric energy over considerable distances have also altered the spatial pattern of manufacturing industries by liberating production from a direct connection to a fixed power system. While in the earlier stages of the industrial revolution, factories located close to sources of energy (a waterfall or a coalfield) or raw materials, mass conveyances, and new energy sources (electricity) enabled much greater locational flexibility.

Industrial development places considerable demands on fossil fuels. At the turn of the twentieth century, the invention and commercial development of the internal combustion engine, notably in transport equipment, made possible the efficient movement of passengers and freight and incited the development of a global trade network. With globalization, transportation is accounting for a growing share of the total amount of energy spent on implementing, operating, and maintaining the international range and scope of human activities.

Energy consumption has a strong correlation with the level of development. Among developed countries, transportation now accounts for between 20 and 25% of the total energy being consumed. The benefits conferred by additional mobility, notably in terms of better comparative advantages and access to resources, have so far compensated the growing amount of energy spent to support this expanded spatial system. At the beginning of the twenty-first century, the transition reached a stage where fossil fuels, such as petroleum, are dominant. Out of the world's total power production, 87% is derived from fossil fuels.

11 Energy and Electricity Mix

As we consider in more detail in this chapter, the breakdown of sources – coal, oil, gas, nuclear, and renewables – is different in the electricity versus the energy mix.[1] Generally, low-carbon sources (nuclear and renewables) account for a larger share in our electricity mix than our total energy mix. This means it is important to distinguish between the two:

- The *energy mix*
- The *electricity mix*

In the beginning, we focus our attention on the *electricity mix*. Where do we get our electricity from? Which countries have the cleanest electricity grids? In this chapter we look at the breakdown across some parts of the world.

Electricity is only one part of total energy – decarbonizing electricity is only one step toward a low-carbon energy system. We must not forget, however, that electrical energy is a conversion energy, i.e., energy in which energy sources are converted into such forms of energy that are useful to humans. Thus, electricity in principle does not exist in nature in such a form that could be used directly, and so whenever we consider electricity, we must take into account the entire conversion cycle – that is to say, we must consider the primary source, from which electricity is obtained. Solar and wind generation are growing quickly across the world. This is, of course, good news as we try to shift our energy systems away from fossil fuels. Such progress often makes the headlines. At first glance, we might think that we are edging close to a fossil-free energy system. Unfortunately, many of these headlines are misleading. Many researchers make the mistake of using the terms *electricity* and *energy* interchangeably, when they are actually not the same thing. *Electricity* (or "power") is just one component of total energy consumption. The other two components being *transport* and *heating*. When we see headlines about our progress on decarbonization, the quoted figures often refer to electricity. Many countries are making progress in clean electricity, but progress in energy as a whole is much slower.

Let us compare the breakdown of the global energy and electricity mix. We can see a large difference between the share that comes from low-carbon sources. Nuclear and renewables account for more than one-third (36.7%) of global electricity. But they account for less than half that figure (15.7%) of the global energy mix. This is because the other elements of the energy demand – transport and heating – rely much more heavily on fossil fuels.

But there is another aspect to consider. The fact that transport and heating are harder to decarbonize, clean electricity will become ever-more important. Many solutions rely on us electrifying other parts of the energy system – such as shifting to

[1] https://ourworldindata.org/energy-mix

electric vehicles. The International Energy Agency, for example, anticipates that by the year 2030, global electricity demand for electric vehicles will have increased 5- to 11-fold compared to levels in 2019. If we are to reap the climate benefits of electric vehicles, this electricity needs to be as low carbon as possible.

But when we see headlines on progress in decarbonizing the electricity sector, we need to remember that it is just one part of the energy story. If we don't, we risk falling into a false sense of progress and let leaders, governments, and companies boast targets that are not nearly ambitious enough.

12 Where Does Our Electricity Come from?

What sources make up our electricity mix? How much comes from coal, oil, and gas, and how much from nuclear, hydropower, solar, or wind? (Fig. 5.4)

The stacked area in Fig. 5.5 shows electricity production in absolute terms. Globally we see that coal, followed by gas, is the largest source of electricity production. Of the low-carbon sources, hydropower and nuclear make the largest contribution, although wind and solar are growing quickly.

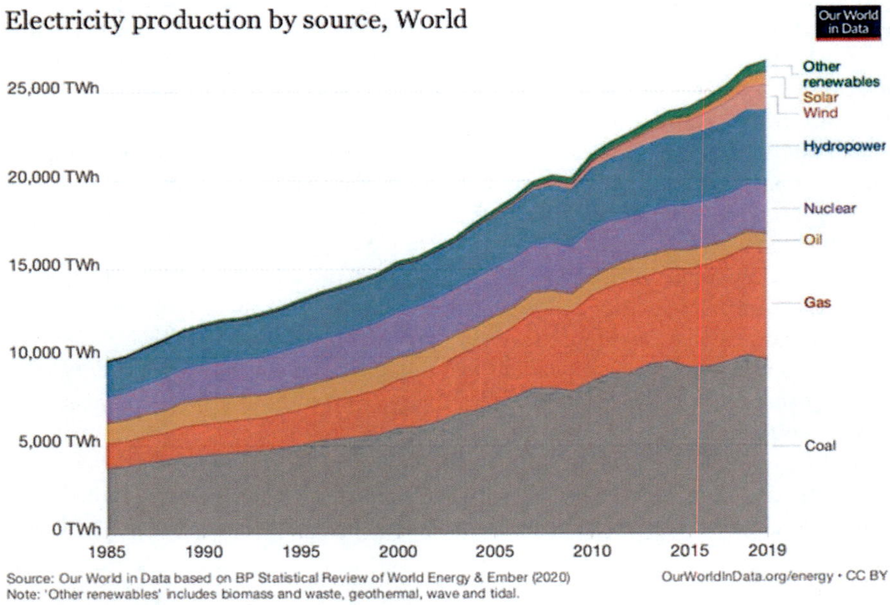

Fig. 5.4 Breakdown of the electricity mix by source (https://ourworldindata.org/electricity-mix)

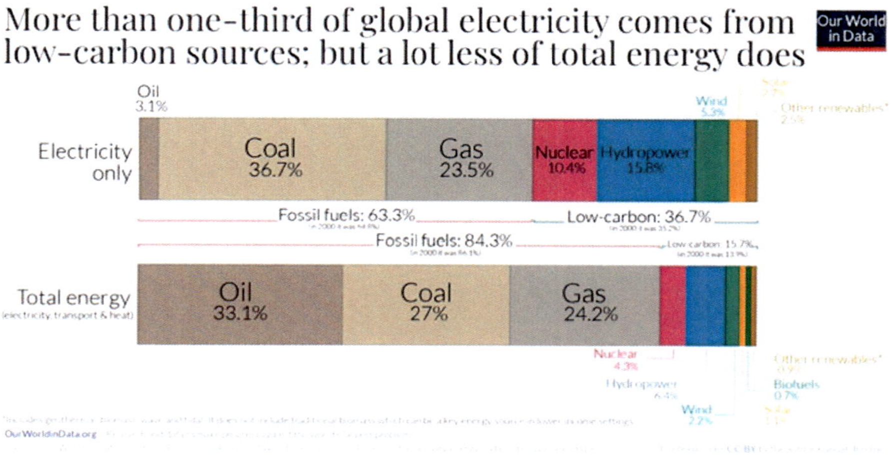

Fig. 5.5 Electricity and energy mix (https://ourworldindata.org/electricity-mix)

13 How Much of Our Electricity Comes from Low-Carbon Sources?

Only around 16% of global *energy* comes from low-carbon sources – that is, the sum of nuclear energy and renewables. But energy and electricity are not the same (see Fig. 5.5). Electricity (sometimes referred to as "power") is only a subset of total energy production, which also includes heating and transport. So, approximately 37% of global electricity production comes from low-carbon sources.

The percentage of global electricity production that comes from nuclear or renewable energy, such as solar, wind, hydropower, and tidal and some biomass, is more than double the share in the total energy mix, where nuclear and renewables only account for 15.7%. When people quote a high number for the share of low-carbon energy in the electricity mix, we need to be aware of the fact that electricity is only part of the energy equation. The share in the total energy mix is much smaller. Disappointingly, the percentage of electricity that comes from low-carbon sources today is almost unchanged from the mid-1980s. In fact, throughout the early-2000s, this share actually regressed mainly because nuclear output declined at a time when renewables have been growing. Nuclear energy has played a key role in low-carbon electricity production for decades. In some countries it is one of – if not, the single – largest source of electricity. For example, France gets more than 70% of its electricity from nuclear power. In some countries we see a dramatic decline in nuclear energy's role as plants have been taken offline. Japan is an obvious example of this, as we see in Fig. 5.6.

Wind accounts for just 5–6% of global electricity production. But wind generation has been growing rapidly in many countries across the world in recent years. Hydropower makes a large contribution to low-carbon electricity across the world. Globally it accounts for around 17% of production.

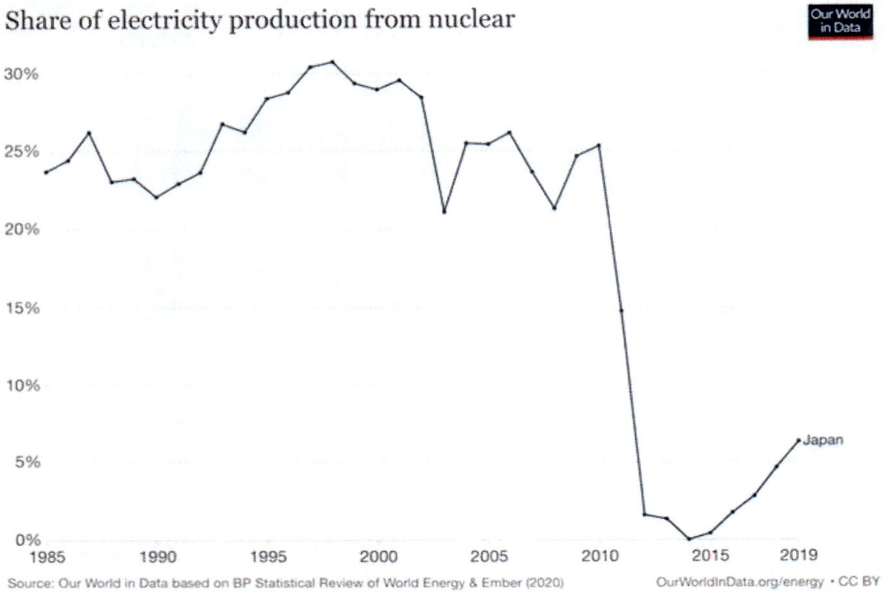

Fig. 5.6 Share of electricity production from nuclear in Japan (https://ourworldindata.org/electricity-mix)

14 Energy Consumption and Sustainable Development

An issue that has raised concerns relates to the capacity of the global economy to accommodate enduring demographic, economic, and resource consumption growth. Population growth and an increase in living standards allow individuals access to an extensive array of goods, energy, and services. Since the 1970s, many statements have been made asserting that the world would be unable to sustain such growth without a possible socioeconomic and environmental breakdown. While these perspectives have been demonstrated to be inaccurate, since resource availability and the quality of life increased, there are concerns that a *threshold* could be reached at some point, particularly regarding climate change. Under such conditions, sustainable development has been advocated as a priority for future social and economic development.[2]

[2] https://transportgeography.org/contents/chapter4/transportation-sustainability-decarbonization/global-sustainable-development/

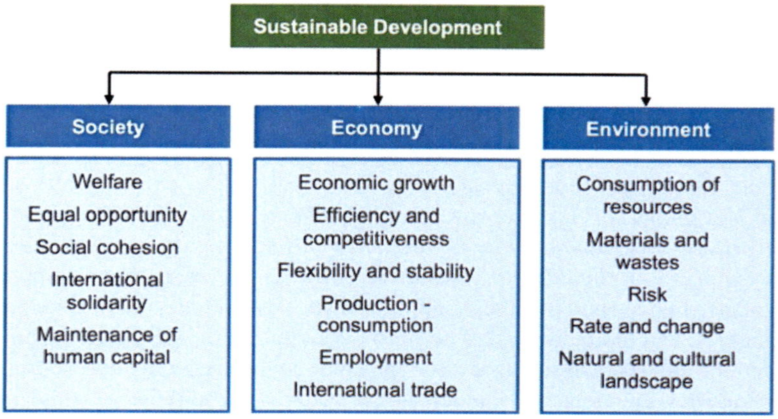

Fig. 5.7 Sustainable development

15 Sustainable Development

Sustainable development is a complex concept that is subject to numerous interpretations since it involves several disciplines and possible interconnections, as show Fig. 5.7. It is not surprising that the subject is prone to confusion regarding its nature, consequences, and appropriate response (Nersesian, 2016). However, it is generally agreed that sustainability favors conditions that benefit the environment, the economy, and the society without compromising the welfare of future generations. Still, as history clearly demonstrates, the conditions of future societies will largely depend upon the legacy of current societies on resources and the environment. All forms of assets (capital, real estate, infrastructures, resources) passed on to the next generation should be at least of equal value (utility) per capita. The basic definition of sustainability has been expanded to include three major pillars:[3]

- *Social equity*. It relates to conditions favoring a distribution of resources among the current generation based upon comparative productivity levels and the promotion of equal opportunities. Social equity is usually the most challenging element of the concept of sustainability to define. It should not be confused with redistribution where a segment of the population agrees or is coerced to support another segment.
- *Economic efficiency*. It concerns conditions permitting higher levels of economic efficiency in terms of resource and labor usage. It focuses on capabilities, competitiveness, flexibility in production, and providing goods and services that supply market demand (Nersesian, 2016).

[3]https://transportgeography.org/contents/chapter4/transportation-sustainability-decarbonization/three-e-development/

- *Environmental responsibility.* It involves an "ecological footprint," which is lesser than the capacity of the environment to accommodate. This includes the supply of resources (food, water, energy, etc.) and the safe disposal of numerous forms of wastes. Its core tenets include the conservation and reuse of products and resources.

Another important debate relates to what extent public entities (both at the national and global levels) have a role to play. There are *competing approaches, one advocating that sustainability be promoted through regulations and the other that the main driver should be individual behavior*. However, both are responsible for the planet's environment – the system (the state), which must provide supportive environments, and the individual, who must act in this spirit and provide their own contribution. Indeed, it seems that environmental advocacy groups are dominantly leaning toward regulations. They would argue that sustainability is a much too long-term concept to be addressed by corporations of individuals focused on the short term. The problem of the whole environment issue depending only on regulations and systems (states) could be in that the time horizon of governments, especially democratic regimes, is also very short, captured by special interests of corporations and powerful individuals, implying that environmental policy is influenced by groups representing *contradictory perspectives*. It is precisely because of this that we need to achieve a global awareness in all individuals, who will:

- Act environmentally friendly and connect into different kinds of environmental protection groups (such as Greenpeace, etc.)
- Put constant pressure on and supervise the decision-makers (governments, corporations, and powerful individuals) in order to create environment-friendly *regulations*

As mentioned on several occasions, environmental issues are not a problem of the individual but a problem of all of us living on this planet, and each of us must contribute their own share in keeping it clean.

The question remains, whether expectations can be placed on entities that seek to optimize positive perception (governments), on entities that seek to optimize efficiency and profit (corporations and powerful individuals), or on individuals. Paradoxically, while governments tend to be inflexible and unable to adapt, corporations have demonstrated a resounding ability to shift their strategies and provide products that reflect customers' expectations (including environmentally responsible products). Further, consumer behavior is a key factor in achieving sustainability as it influences the provision and the delivery of goods. It could thus be argued that the private sector is more likely to achieve sustainability than the public sector, particularly since the benefits are apparent. However, just like governments, the private sector also requires some kind of control (supervision), for which we are all responsible. This complex relationship underlines the respective roles of regulations, innovations, and control in achieving a higher level of sustainability (Nersesian, 2016).

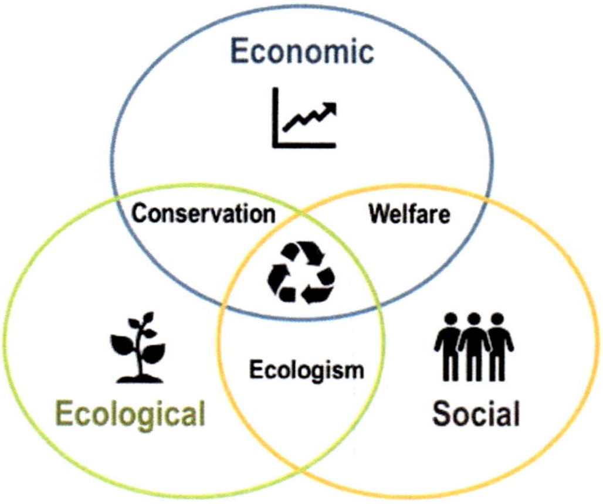

Fig. 5.8 Global sustainability

Societies do not contribute to environmental problems at the same level. As we point out, sustainability can be thus expressed at two spatial levels:

- *Global.* Long-term stability of the Earth's environment and availability of resources to support human activities (see Fig. 5.8).
- *Local.* Localized forms are often related to urban areas in terms of jobs, housing, and environmental pollution.

16 Conclusion

As we saw in this chapter, the breakdown of sources – coal, oil, gas, nuclear, and renewables – is different in the electricity versus the energy mix. Generally, low-carbon sources (nuclear and renewables) account for a larger share in our electricity mix than our total energy mix. This means it is important to distinguish between the two.

The main premise here is that electricity is a clean energy. But questions arise: What sources constitute our electricity mix? How much comes from coal, oil, and gas, and how much from nuclear, hydropower, solar, or wind? The answers to these questions are related to the idea of sustainability. However, sustainable development is a complex concept that is subject to numerous interpretations since it involves several disciplines and possible interconnections. It is not surprising that the subject is prone to confusion regarding its nature, consequences, and appropriate response. It is generally agreed that sustainability favors conditions that benefit the environment, the economy, and the society without compromising the welfare of future generations. Still, as history clearly demonstrates, the conditions of future societies will largely depend upon the legacy of current societies on resources and the environment. All forms of assets (capital, real estate, infrastructures, resources) passed on to the next generation should be at least of equal value (utility) per capita. The basic definition of sustainability must be expanded to the three major pillars:

- *Social equity.*
- *Economic efficiency.*
- *Environmental responsibility.* This involves an "ecological footprint," which is lesser than the capacity of the environment to accommodate. It includes the supply of resources (food, water, energy, etc.) and the safe disposal of numerous forms of wastes. Its core tenets include the conservation and reuse of products and resources.

As mentioned on several occasions, environmental issues are not a problem of the individual but a problem of all of us living on this planet, and each of us must contribute their own share to keeping it clean.

Chapter 6
Green Solutions

Abstract This chapter, *Green Solutions*, discusses various forms of energy extraction and production and points out the need for a rational use of energy. In order to be deemed green energy, a resource cannot produce pollution, such as is found with fossil fuels. In this chapter, the current situation on greenhouse gas emissions by pollutant, and by sector, is presented.

Greenhouse gas emission in the world by pollutant (Source: UNFCCD Interface)	Greenhouse gas emission in the EU by sectors (Source: UNFCCD Interface)
Carbon dioxide (CO_2) – 81% Methane (CH_4) – 11% Nitrous oxide (N_2O) – 5% Hydrofluorocarbon (HFCs) – 2%	Energy – 80.70% Industry – 7.82% Agriculture – 8.72% Waste management – 2.75%

If we want to address and solve the problem of energy production and consumption and, at the same time, to reduce greenhouse gas resources and act green, there are only two diametrically opposed options, saving energy or using green energy sources. In between the two options, there is a range of intermediate, hybrid solutions, including more or less intense energy saving and the use of more or less environmentally friendly energy sources. Of course, modern technologies also play an important role in this respect, as they enable the reduction of greenhouse gas emissions through higher efficiency and cleaning of toxic gas emissions, which helps to partially neutralize their harmful impact on the environment. This can be achieved through a combination of general regulations at a global level and raising the awareness of individuals.

Let us consider the current situation on greenhouse gas emissions by pollutant and by sector.

Greenhouse gas emission in the world by pollutant (Source: UNFCCD Interface)	Greenhouse gas emission in the EU by sectors (Source: UNFCCD Interface)
Carbon dioxide (CO_2) – 81% Methane (CH_4) – 11% Nitrous oxide (N_2O) – 5% Hydrofluorocarbon (HFCs) – 2%	Energy – 80.70% Industry – 7.82% Agriculture – 8.72% Waste management – 2.75%

B. Aberšek, A. Flogie, *Human Awareness, Energy and Environmental Attitudes*,
https://doi.org/10.1007/978-3-031-05871-4_6

If we want to address and solve the problem of energy production and consumption and, at the same time, to reduce greenhouse gas resources and act green, there are only two diametrically opposed options:

- Saving energy
- Using green energy sources

In between the two options, there is a range of intermediate, hybrid solutions, including more or less intense energy saving and the use of more or less environmentally friendly energy sources. Of course, modern technologies also play an important role in this respect, as they enable the reduction of greenhouse gas emissions through higher efficiency and cleaning of toxic gas emissions, which helps to partially neutralize their harmful impact on the environment. As mentioned, this can be achieved through a combination of general regulations at a global level and raising the awareness of individuals.

1 Rational Use of Energy

There are more than enough reasons why we should save energy! What are the next steps in this direction? World energy consumption is already off the charts. This is not an exaggeration: it is a fact that in the last 150 years, energy consumption has increased by as much as 350 times. Most of the world's energy, about 90%, is obtained from fossil fuels (coal, oil, gas), which have formed over hundreds of millions of years. Unless we limit their consumption in time, we will soon run out of these resources completely. Gas and liquid fuels are expected to run out as early as the next century. While there may be more coal reserves, these could still not last for a very long time. In addition, they are environmentally problematical and unacceptable according to today's standards. Even though we are well aware of this, we continue to behave carelessly and irresponsibly. In addition, we have no way of knowing in what way and with what risks the future generations will obtain energy.

1.1 Incinerating Precious Raw Materials

Fossil fuels are valuable raw materials. We obtain a number of basic chemical compounds from them. Many industries require raw materials in order to produce a wide variety of plastics and products, without which it would be difficult to imagine our lives.

1.2 Threats to the Environment

When fossil fuels are burned, many harmful substances are released, which seriously poison the atmosphere in populated areas and especially in the wider surroundings of large industrial facilities and thermal power plants. The most harmful gases are CO, CO_2, SO_2, nitrogen oxides, and smoke particles. The emission of these harmful substances can be reduced in various ways. The most useful thing to do, however, would be to reduce energy consumption.

1.3 Excessive Emissions of Nonpoisonous Gases Are Also Dangerous

CO_2 and other greenhouse gas emissions are among the most pressing environmental problems. Even nontoxic gases, however, act like greenhouse glass. They let in sunlight and allow the Earth to warm up, thus affecting the Earth's heat radiation and preventing the cooling of the atmosphere. When 1 kilogram of quality brown coal burns, more than a kilogram of CO_2 is emitted, but when 1 kilogram of liquid fuel burns, as much as 2.8 kg of CO_2 is produced. With the massive consumption of fossil fuels around the world, huge amounts of CO_2 are released, already exceeding 20 billion tons per year. The consequences are certainly not negligible. In recent years, the concentration of CO_2 in the air has increased by more than 25%. This increased the global greenhouse effect, and the atmosphere is now already warmer by more than 1 degree Celsius.

Warming in itself is not a threat to humanity. The danger lies in the consequences: unstable weather conditions, severe droughts, and periods of storms and hurricanes associated with floods and erosion processes. The melting of glaciers, which causes the oceans to rise, is extremely unfavorable. Extensive working and cultivation areas face the threat of becoming submerged. It is expected that the effects of doubling the concentration of CO_2 in the Earth's atmosphere will be seen around the year 2030. This is, therefore, not some distant future but a time that we ourselves will live in. Unfortunately, cleaning plants alone will not be able to reduce CO_2 emissions. This can be achieved by reducing energy consumption or through some completely new, revolutionary technology.

The most logical (but also dystopian) solution is *saving energy*.

The stress on the environment, therefore, can be reduced by three sets of measures:

- *Energy saving*
- *Use of the latest cleaning technologies*
- *Use of alternative energy sources*

We ourselves need to make the decision to save energy!

1.3.1 Saving Energy at the Level of the Individual

Various kinds of energy-saving heating devices and thermal protection substances have been developed so far, but it is still a choice of every individual, how they are going to save energy.

Cold air may be entering your apartment from the staircase through a leaky door. This can be prevented by attaching a seal to the bottom edge of the door. Placing an airbag on the outside of the window blinds or shutters overnight will help reduce energy consumption by as much as 15%.

Ventilation and heat loss. In order to feel good, we need fresh air and, therefore, ventilation. Letting the air flow through a room in an inappropriate way will consume as much as 20% more energy. The correct way to ventilate is to close the valve on the radiator and wide open the window for a few minutes. We need to repeat this in all rooms three times a day. The colder the temperature outside, the shorter the ventilation time. Of course, there are various modern technological solutions at our disposal today, such as heat recovery (heat recuperation), whereby the warm outgoing air emits its heat to the cold, incoming air.

Heat consumption depends significantly on the temperature of the room. Two degrees of difference (between 20 and 22 °C) may not seem a lot at first glance, but they mean up to 12% of additional heat consumption. Because of this, it is very economical to install control devices, such as thermostatic radiator valves. State-of-the-art dual-stage thermostats allow you to set the daytime and nighttime temperatures separately. Heating should not be turned off completely during the night. Reheating requires much more energy in comparison to constantly maintaining a certain temperature. With proper settings and regulation, you can save up to 20% of energy. There are many more different solutions for saving energy out there; some are based on common sense, while others are highly technological.

2 Back to the Future: Possible Green Solutions

No artificial way of producing energy is entirely environmentally friendly. Only nature knows how to create energy in a way unharmful to herself.

The most sensible solution for the modern age is to use more environmentally friendly energy sources, which we may refer to as green solutions. As indicated earlier, the current stage of technological development does not offer a perfect solution for green energy production. Existing ways of energy extraction are all more or less harmful to the environment. It could be argued that only nature knows how to create energy in a way unharmful to herself.

Forecasts suggest that in a hundred years' time, energy from nonrenewable energy sources will be enough to meet the needs of about one-third of the planet's population, i.e., 4 billion people out of 12. About eight billion, however, will be dependent on the energy sources of the future, e.g., nuclear fusion reactors. Fewer

Table 6.1 Characteristics of different energy sources

Energy source	Capital costs (€/kW)	Production costs (€/kWh)	Average plant power (kW)	Harmful gas emissions (g/kWh)	Required surface (km^2/1000 MW)
Oil products	1000	0.25	$1–10^3$	200	1
Coal	800–1100	0.05	$10^3–10^6$	270	~1–2.5
Gas	300–600	0.035–0.050	$10^3–10^6$	180	1
Nuclear fission	100–1500	0.05–0.08	$2.5*10^5–10^6$	6	~1
Hydropower plants	1400	0.05	$10^5–2*10^7$	20	30–40
Solar energy	400–6000	0.25	$10^{-2}–10$	25	23
Wind energy	700–1200	0.06–0.10	$10^{-1}–10^5$	34	490
Biomass	1300–1700	0.05–0.10	$1–15*10^4$	10	2000
Fusion energy	6000	0.05–0.10	$10^6–3*10^6$	9	1

Nuclear, solar, wind, and fusion energies generate CO_2 during construction and decomposition. Depending on the capacity factor, the total cost can increase several times, even up to ten times (e.g., solar energy). Fusion energy is expected to be commercially usable after 2050; the estimated costs are calculated for that period in the table below. In anticipating the required surface, one needs to consider safety zones, which occupy a considerably large space (e.g., nuclear fission)

and fewer people will depend on the energy produced by fission reactors, oil, and natural gas. Based on research, predictions for the future include the following:

- In the next 30 years, the amount of energy from fossil fuels will be the same as the amount used over the entire course of human history.
- Fossil energy sources will continue to shape energy supply.
- Renewable energy sources such as biomass, wind, solar, and water will become increasingly important.
- Nuclear energy will become very important in the future.
- Energy policies will affect environmental protection measures.

Only new energy sources will be able to ensure the supply of energy in the future. Among the most commonly used new energy sources, there will be solar radiation, kinetic energy produced by wind, nuclear fission of heavy chemical elements, and fusion of light element nuclei (nuclear fusion) (Table 6.1).

It is expected that the world population will increase drastically in the current century. According to some studies, the number of people will increase to 10 billion by the middle of the twenty-first century (for more details, see Table 3.1 and Fig. 3.3). The share of the population in developing regions contributes to higher energy consumption, as access to energy increases the quality of life. In addition, most countries seek independence from the world's largest oil producers and are therefore looking for alternative energy supply options (Fig. 6.1).

For an individual person, the average energy consumption in 1990 was 2.2 times 10^{11} J (in physics, joule is a unit of work and energy, named after the English

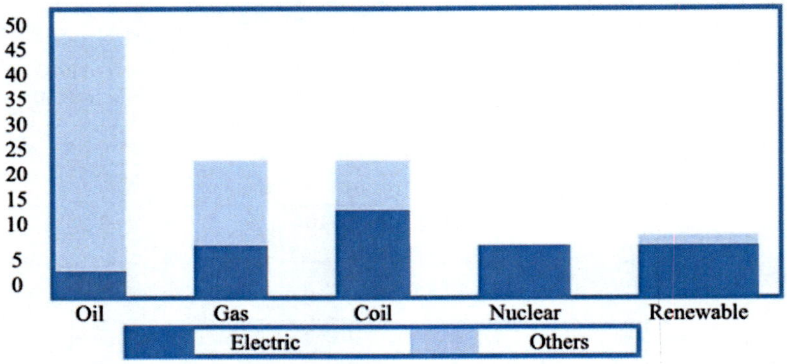

Fig. 6.1 The share of sources of global energy consumption (electricity and other forms)

Table 6.2 The growth of energy demand according to two scenarios: standard growth and reduced growth

	By inhabitant (T oil/year)			Global needs (10^9 T oil/year)		
		2050			2050	
Group of countries	1988	Standard growth	Reduced growth	1988	Standard growth	Reduced growth
OECD[a]	5.2	5.2	2.6	4.0	4.6	2.3
Central and Eastern European states	4.4	4.4	2.2	1.9	2.1	1.1
Developing countries	0.5	1.5	1.0	2.0	13.8	9.2
World (total)	1.5	2.0	1.2	7.9	20.5	12.6

Tons of oil equivalent represents the amount of energy that can be obtained from one ton of oil
[a]*OECD* (Organization for Economic Co-operation and Development) brings together 37 member states committed to democracy and a market economy. The OECD is considered an organization of elite economies, as OECD member countries generate more than half of the world's GDP with less than a fifth of the world's population

physicist James Prescott Joule). By the year 2050, according to different scenarios (see Table 6.2 below), primary energy consumption will be 2–3 times higher than today.

Reports of the Intergovernmental Panel on Climate Change (IPCC)[1] show that greenhouse gas emissions over the last century have led to rising temperatures, affecting weather patterns. In order to stabilize the concentration of greenhouse gases in the atmosphere, humanity must adapt its energy production methods accordingly.

Today, global energy needs are largely covered by fossil fuels. If conventional oil and gas resources are not used more efficiently in the future, they will only be sufficient for the next 60 years. Oil is currently the main source of energy needed for

[1]Intergovernmental Panel on Climate Change (IPCC).

transport, and air transport has no alternative to oil at all. Moreover, certain alternatives, such as fuel cells, also require electricity to operate in the first place. Renewable energy sources (solar energy, water, wind, biomass, geothermal energy) provide certain solutions, but in developing countries, they cannot cover all energy needs. We may not want to admit it, but the only current (short-term?) solution is nuclear energy, perhaps not nuclear energy in the way we understand it today but a modification of it; let us refer to it as *nuclear energy 4.0*. Nevertheless, the physics behind the extraction methods remains the same.

3 Low-Carbon Solutions

3.1 Nuclear Fission

Nuclear fission is a physical reaction in which a heavy atomic nucleus[2] splits into two medium-heavy nuclei, emitting neutrons.[3] In heavy nuclei such as uranium, fission can be released spontaneously, or it can be caused by a neutron or photon that is absorbed in the nucleus and transmits energy to it. The greater this transmitted energy, the more stable the nuclei that can be divided. Elements from the end of the periodic table (uranium, thorium, plutonium) are best suited for neutron absorption. Uranium and its isotope ^{235}U are the most suitable for technical use. After the neutron is captured, it splits into two medium-sized nuclei, as shown in Fig. 6.2. Two to three neutrons are emitted during this nuclear fission.

New nuclei have an excess of neutrons, so they quickly decay into more stable nuclei, emitting electrons, photons, and possibly neutrons.

An example of fission:

$$^{1}n + ^{235}U \rightarrow ^{236}U^{*} \rightarrow ^{98}Zr + ^{136}Te + 2^{1}n. \tag{6.1}$$

Of course, ^{98}Zr and ^{136}Te are also radioactive and decay further. Cases when at least 200 MeV of binding energy is released at the cleavage are considered to be useful decays. Approximately 170 MeV is released in the form of kinetic energy of fission nuclei, 10 MeV in the form of kinetic energy of generated neutrons and photons, and the rest is represented by the energy of electrons, photons, and delayed neutrons emitted by nuclei during radioactive decay.

---

[2] The atomic nucleus is the central part of the atom, representing more than 99.9% of the total mass of the atom. The diameter of an atom is about 10^{-15} m in size and depends on the individual atom. The nucleus is made up of nucleons and neutrons. Nucleons are positively charged protons and neutrons are uncharged. Nucleons are interconnected by a strong nuclear force.

[3] A neutron is a nuclear particle without an electric charge and with a mass of $1675 \cdot 10^{-27}$ kg. Neutrons together with similarly heavy protons make up atomic nuclei.

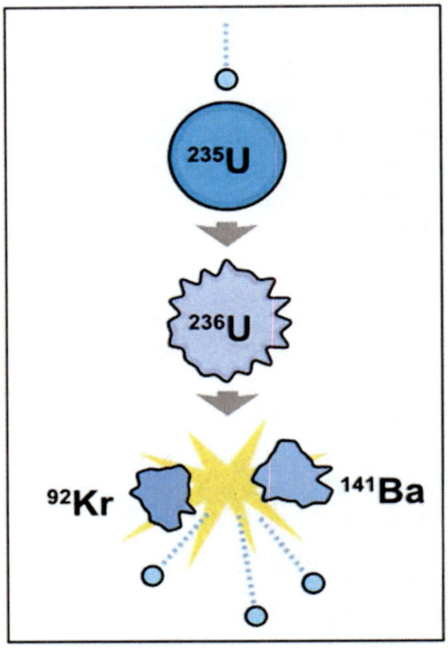

Fig. 6.2 Uranium fission

The neutrons released during the fission can cause new fissures in the presence of a larger amount of the ^{238}U isotope. This is referred to as a nuclear chain reaction (Fig. 6.3).

If the chain reaction is uncontrolled, the number of splits increases greatly from generation to generation, ending in an explosion, as huge amounts of energy are released in a very short time. An example of such an uncontrolled chain reaction is the atomic bomb. Of course, the fission of atoms for safe energy production in nuclear reactors is controlled.

3.2 Types of Nuclear Fission Power Plants Today

The specificity of nuclear power plants is in the way they exploit the processes inside the reactor, as a source of thermal energy. An entire nuclear power plant consists of the following:

- Primary components
- Secondary and tertiary components

The primary components include the reactor, the elements of the primary circuit, and the auxiliary components of the reactor. The secondary components are the turbine and electrical mechanisms of the power plant, and the tertiary components are mostly parts of the cooling system. The secondary parts of fission nuclear power

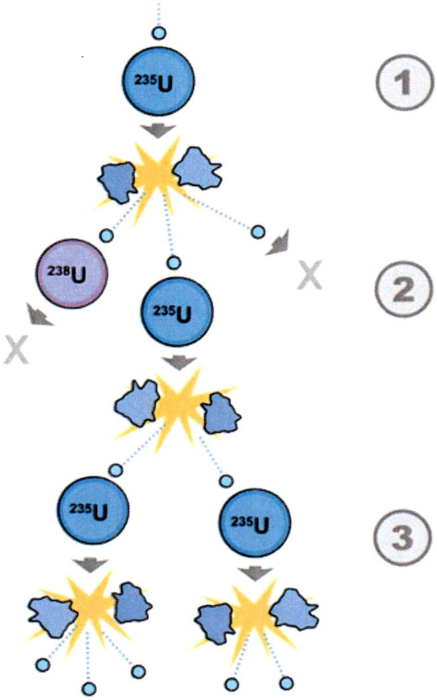

Fig. 6.3 Nuclear chain reaction

plants are similar to those of conventional thermal power plants. According to the composition of primary mechanisms, we distinguish between different types of fission nuclear power plants today:

- The BWR[4] is a nuclear power plant with water-moderated and cooled reactor systems. The water evaporates in the reactor and the steam then drives the turbine. A BWR is similar to a thermal power plant, except that the boiler is replaced by a reactor.
- The PWR[5] (pressurized water reactor) is moderated and cooled with ordinary water (Fig. 6.4). The cooling water is kept under pressure so that it cannot boil. In the secondary loop, water boils and evaporates. Most power plants are built according to this model, including the Slovenian nuclear power plant Krško (Fig. 6.4).
- The PHWR[6] uses heavy water as its coolant and neutron moderator. The heavy-water coolant is kept under pressure to avoid boiling the steam for the turbine is

[4]BWR – boiling water reactor.

[5]PWR – pressurized water reactor.

[6]PHWR – pressurized heavy-water reactor.

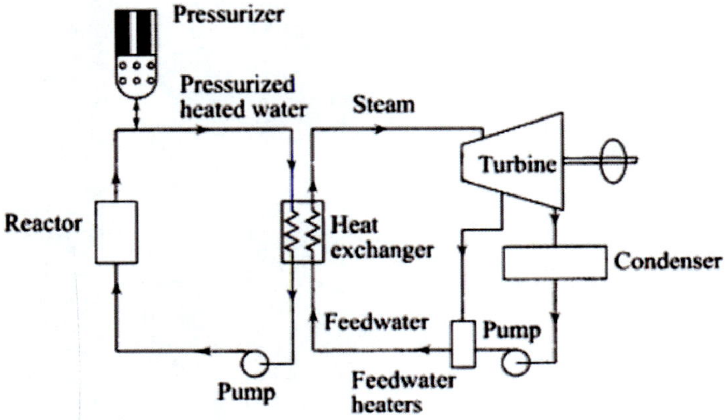

Fig. 6.4 Schematic diagram of a PWR

produced in a secondary circuit. These power plants are also known as CANDU (Canada Deuterium Uranium).

- A GCR[7] is a nuclear reactor that uses graphite as a neutron moderator and a gas (carbon dioxide or helium) as coolant. The gas heats the water, which evaporates and drives the turbine. Most of GCR power plants operate in the United Kingdom.
- A light-water graphite reactor LWGR[8] uses water as coolant. The water in the reactor boils, after which the steam is separated from the water droplets in a steam separator.
- In addition to the reactors in nuclear power plants, there are around 300 research, experimental, and school reactors worldwide. Around 500 nuclear propulsion systems are also installed on vessels around the world.

The design and technology of reactors is constantly upgraded and elaborated. Some of the advantages that newer reactors have in comparison to older ones are as follows:

- New reactors require less time to be constructed and installed.
- Newer reactors are more affordable.
- Newer reactors are safer (it is ten times less likely that the reactor core will be damaged).
- Newer reactors produce less low-level radioactive and medium-level radioactive waste.

Nuclear fuel used in fission is any material that can be used to generate nuclear energy. Nuclear reactors most often use heavy radioactive elements, which

---

[7]GCR – gas-cooled reactor.

[8]LWGR – light-water-cooled graphite-moderated reactor.

themselves maintain a controlled chain reaction inside the reactors. In most cases, ^{238}U and ^{239}Pu are used. Uranium, as the most often used element for the production of nuclear energy, is extracted from the uranium ore, phosphorites, coal ash, seawater, and granite.

3.2.1 Generation IV Reactor Concepts

In recent decades, significant advances have been made in the development of fission nuclear reactors, and so today we are talking about *generation IV nuclear reactor systems* (Schulenberg et al., 2004):

I. *Gas-Cooled Fast Reactor (GFR)*

The GFR system is a helium-cooled reactor with fast-neutron spectrum and a closed fuel cycle, which is primarily envisioned for electricity production and actinoide (actinide) management. Using the breeder concept improves the use of the nuclear fuel by two orders of magnitude compared to current thermal reactors.

II. *Lead-Cooled Fast Reactor (LFR)*

The LFR systems are reactors cooled by liquid metal (lead or lead/bismuth) with a fast-neutron spectrum, a closed fuel cycle, and a wide range of unit sizes from small "batteries" up to large single plants. The LFR battery option is a small factory-built turnkey plant with a very long core life (10–30 years). With fast neutrons and the closed fuel cycle, an efficient conversion of fertile uranium and the use of actinoides are achieved.

III. *Molten Salt Reactor (MSR)*

The MSR system, primarily envisioned for electricity production and waste burndown, is based on a thermal neutron spectrum and a closed fuel cycle, where the uranium fuel is dissolved in the sodium fluoride salt coolant that circulates through graphite core channels. Fuel loading, reprocessing, and separation of fission products during operation enable a high availability. The reactor also serves to eliminate actinoides by simply adding them to the molten salt.

IV. *Sodium-Cooled Fast Reactor (SFR)*

The SFR system consists of a fast-neutron reactor and a closed fuel cycle. This reactor type mainly serves to eliminate highly radioactive waste, plutonium, and other actinoides.

V. *Supercritical-Water-Cooled Reactor (SCWR)*

The SCWRs are high-temperature, high-pressure water-cooled reactors that operate at pressures and temperatures at which there is no difference between liquid and vapor phases (permitting to save expenses for components like heat exchangers). This reactor type was designed mainly to generate cheap electricity.

VI. *Very High-Temperature Reactor (VHTR)*

The VHTR system uses a thermal neutron spectrum and a once-through uranium fuel concept. The main purpose of this gas-cooled reactor type is the generation of power, hydrogen, and process heat.

In 2002, the Technology Roadmap for Generation IV Nuclear Energy Systems was published and updated in the 2014. All member states of this road map used it as basis to prepare and conduct their research and development programs (NERAC, 2002). While research into the individual systems is being performed independently of each other, other problem areas are to be solved together for all six systems (e.g., closed fuel cycle, development of fissile materials and material features, hydrogen production, safety and reliability, economic efficiency, physical protection, and proliferation barriers).

4 Nuclear Fusion

There is no way we could imagine life today without the use of energy. The fossil fuel reserves on Earth are severely limited. In the future, humanity will use up most of its primary energy reserves, which can be exploited economically today, but new, renewable sources will certainly not be able to compensate for the loss of these energy sources and meet the growing energy needs of the planet's population. The next generations of humans will face a shortage of natural gas, coal, and uranium isotopes, which are currently satisfying our energy needs. These resources will be replaced by others, which are used today to a much lesser extent.

Many of the new energy sources are as effective as they are fascinating, for example, the still relatively unexplored and untested nuclear fusion. In recent years, scientists have explored the possibilities of obtaining energy through fusion only to a lesser extent, with smaller projects. The first major project is being carried out today in France, where an experimental fusion reactor with the project name ITER is being built and tested. With the knowledge gained through the ITER project experience, it will be possible to set up the first commercial fusion power plants later. Let us now consider the basics of the energy-producing mechanisms in a nuclear fusion reactor.

5 Fusion Energy

All operating nuclear power plants today are based on the principle of fission of heavy nuclei (uranium and plutonium). Fission nuclear power plants are technologically advanced enough, so they are relatively easy to build, economically viable, and technologically quite sophisticated. However, fission energy and its extraction have certain disadvantages, such as:

- Limited access to raw materials for fuel.
- Potential hazard due to large amount of radioactive material.

– Larger amount of radioactive waste (spent fuel); the storage of radioactive waste is technologically demanding and a huge stress to the environment.

However, there is another important way of harnessing nuclear energy, nuclear fusion (E. C., 2004). Nuclear fusion was discovered even before the discovery of the fission process. English physicists John Cockroft and Ernest Walton first demonstrated the display of fusion energy in 1934. In 1936, Ernest Rutherford anticipated that a nuclear fusion reaction between deuterium and tritium would yield the best results.

The invention of the hydrogen bomb[9] in the early 1950s is considered to be the first uncontrolled use of fusion energy in the world. However, since the very beginnings, when the first discoveries were made in the field of nuclear energy, scientists have been thinking about *controlled fusion*. By using fusion as an energy source in a technologically and economically acceptable way, we would be able to solve the problems of energy supply for the entire planet in virtually unlimited quantities. Energy solutions in the field of nuclear fusion would certainly also improve the safety of nuclear power plants.

However, advances made in the way we exploit nuclear fusion as an energy source are still hindered by certain technical difficulties. Due to this, there is still no certainty today as to the exact time when we are going to be able to use fusion energy for commercial purposes. All current research, as well as future research for the next 20 years, is based on the knowledge and solutions of *today's* scientists and their level of thinking about physics. The main issues are two: pressure and temperature (at least 100 million K). Scientists have tried to circumvent these two fundamental problems with more or less utopian ideas, one of which is known as *cold fusion*.

5.1 Cold Fusion

Cold fusion,[10] also called low-energy nuclear reactions (LENR) or condensed matter nuclear science, is a nuclear reaction that would hypothetically occur at room temperature and room pressure (Huizenga, 1993).

In 1989, Martin Fleischmann and Stanley Pons reported a "laboratory" nuclear fusion reaction that had been successfully tested at the laboratory of the University of Utah. They documented an anomaly of excess temperature using electrolytic cells in

[9] A hydrogen bomb (also known as a thermonuclear bomb, H-bomb, or fusion bomb) is a type of nuclear weapon that exploits the large amount of nuclear energy released during the process of nuclear fusion. Hydrogen bombs use the Teller-Ulam configuration, in which a hydrogen bomb is detonated in a specially designed area that is in contact with the hydrogen fuel. The gamma and X-rays of the explosion compress and heat a capsule of tritium, deuterium, or lithium deuteride starting a fusion reaction. Neutrons emitted by this fusion reaction can induce a final fission stage in a depleted uranium tamper surrounding the fusion fuel.

[10] https://theworld.com/~mica/cft.html

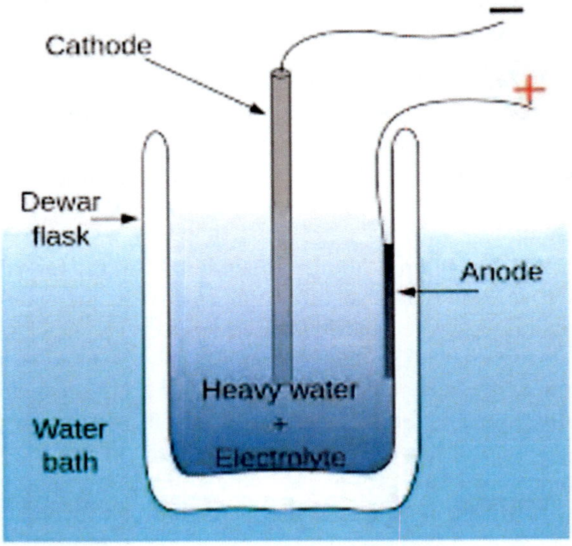

Fig. 6.5 Simplified schema of an electrolytic cell, such as the one used by Fleischmann and Pons in their experiment

the electrolysis of heavy water using palladium (Pd) electrodes (Fig. 6.5). In searching for the cause, they put forward the dubious hypothesis that the abnormal change in temperature was caused by nuclear fusion of deuterium (D). With this conclusion, they raised hopes that a cheap source of energy in infinite quantities had been discovered.[11]

In their experiment, Fleischmann and Pons used a double-walled vacuum chamber (Fig. 6.6), reducing the temperature losses to 5%. The chamber was then subsequently immersed in a constant temperature solution to reduce external influences. The tank was partially open to release gasified deuterium and oxygen, and heavy water had to be supplied to the system at intervals. During the experiment, special attention was also paid to the purity of the palladium electrode and electrolyte, as changes in the structure of the material could occur during a long experiment. The experiment was equipped with a temperature meter, which could measure the temperature of the electrolytes, and an electric heater, which supplied the lost heat to the heated gas. The results of the experiment showed certain increases in temperature from the standard 30 ° C to elevated 50 ° C. The temperature would reach a constant value for several weeks and then an inexplicably elevated temperature.

In the years to follow, however, cold fusion was deemed a "pathological science" due to subsequent unsuccessful replications of the Fleischmann-Pons experiment. In 1989, the US Department of Energy finally refuted their theory. Nonetheless, cold fusion later remained the starting point for several scientific discussions, even though their findings were received with much skepticism among most scientists.

[11] https://undsci.berkeley.edu/article/cold_fusion_01

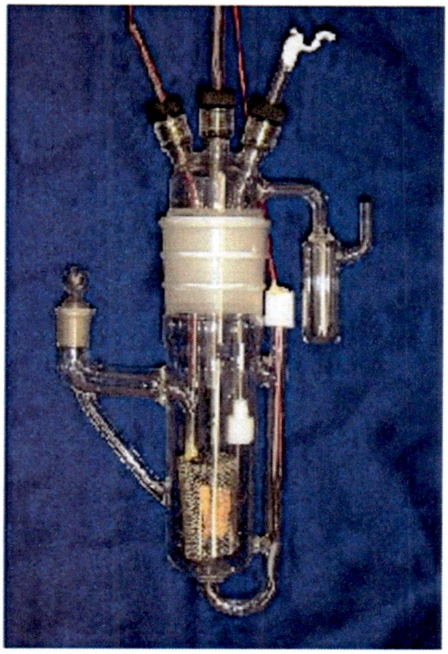

Fig. 6.6 Experimental cell for a possible use in cold fusion (US Navy Space and Naval Warfare Systems Center, San Diego). (https://en.wikipedia.org/wiki/Cold_fusion)

5.2 *From Utopia to the Physical Basics of Fusion*

Analyses of the stability of atomic nuclei and the dependence of the energy of the nucleon bond on the mass number can be generalized to conclude that nuclear energy can be released only by nuclear reactions leading to the formation of atomic nuclei with higher nucleon bond energy (Jones et al., 1989). Nuclei with the highest energy of nuclear bonds, i.e., the most stable nuclei, are those with a mass number of around 60.

The generation of energy occurs with those nuclear reactions, which result in nuclei, and whose mass numbers are closer to the most stable nuclei. These are as follows:

- Nuclear fission reactions of heavy nuclides
- Nuclear fusion reactions of light nuclides

Fission was discussed already in the previous chapter; let us focus here only on nuclear fusion reactions of light nuclei. In the fusion of atomic nuclei, weaker-bound light nuclei merge into stronger-bound heavier nuclei. This process releases a certain amount of energy. The mass of a stable nucleus is less than the sum of the masses of the protons and neutrons involved. The difference between the sum of the masses of all the nucleons that make up the nucleus and the mass of the nucleus is called the mass defect:

$$\Delta m = Zm_p + (A - Z)m_n - m_z \tag{6.2}$$

whereby m_z marks the mass of the nucleus, m_p marks the mass of the proton, and m_n marks the mass of the neutron.

When a new nucleus is formed, the mass defect (Δm) is converted into the binding energy of the nucleus according to Einstein's formula:

$$E = m_z c^2. \tag{6.3}$$

If the mass defect, measured in energy units, is divided by the number of nucleons in the nucleus, the binding energy of the nucleon is obtained. The greater this energy, the stronger the bonds in the nucleus, and consequently the nucleus is bound more strongly. Whenever weakly bound nuclei are formed in nuclear reactions to form more strongly bound nuclei, the difference in binding energies is released.

The nucleus of atoms cannot simply be forced into a fusion. The nuclei of atoms are positively charged, so they repel each other due to the coulomb force. They must be placed close enough to each other, so that the attraction of the nuclear force can overcome the Coulombian repulsive force.

The basic materials for fusion nuclear reactions are deuterium (D) and tritium (T). Deuterium is extracted from water using methods similar to those developed for heavy-water reactors. Tritium is obtained through one of the following nuclear reactions of a neutron with lithium (Li):

$$\begin{aligned} &{}^{7}\mathrm{Li} + n(\text{fast neutron}) \rightarrow \mathrm{T} + {}^{4}\mathrm{He} + n, \\ &{}^{6}\mathrm{Li} + n(\text{thermal neutron}) \rightarrow \mathrm{T} + {}^{4}\mathrm{He} + 4,8\ \mathrm{MeV}. \end{aligned} \tag{6.4}$$

The fast neutrons produced in the DT reaction react with ^{7}Li. The neutrons resulting from this reaction are slow, so they can react with ^{6}Li. Tritium can be obtained by means of a coat of lithium salt solution surrounding the fusion chamber. By cooling this lithium coating, energy is obtained in the form of heat that can be dissipated from the system. This heat can later be used as useful energy for other purposes (Fig. 6.7).

Thus, fusion nuclear reactions are suitable to be applied for energy production in the future:

$$\begin{gathered} \mathrm{D} + \mathrm{T} \rightarrow {}^{4}\mathrm{He} + n + 17,6\ \mathrm{MeV}; \\ \mathrm{D} + \mathrm{D} \rightarrow {}^{3}\mathrm{He} + n + 3,2\ \mathrm{MeV}; \\ \mathrm{D} + \mathrm{D} \rightarrow \mathrm{T} + {}^{1}\mathrm{H} + 4,0\ \mathrm{MeV}; \\ \mathrm{D} + {}^{3}\mathrm{He} \rightarrow {}^{4}\mathrm{He} + {}^{1}\mathrm{H} + 18,3\ \mathrm{MeV}; \\ {}^{6}\mathrm{Li} + n \rightarrow {}^{4}\mathrm{He} + \mathrm{T} + 4,8\ \mathrm{MeV}; \\ {}^{7}\mathrm{Li} + n(+2,5\ \mathrm{MeV}) \rightarrow \mathrm{T} + {}^{4}\mathrm{He} + n; \\ {}^{11}\mathrm{B} + {}^{1}\mathrm{H} \rightarrow 3{}^{4}\mathrm{He} + 8,7\ \mathrm{MeV}. \end{gathered} \tag{6.5}$$

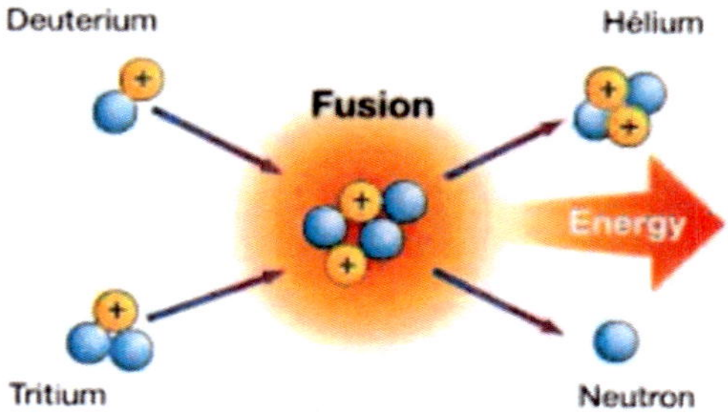

Fig. 6.7 The fusion of deuterium and tritium produces helium, a neutron, and a large amount of released energy

The energies released in certain reactions range from 3.2 to 18.3 MeV and are shown in Eq. 6.5. For example, in the DT reaction, the energy released at a single fusion of D and T nuclei is as much as 17.6 MeV. Its products are the helium nucleus (He) and the neutron (n).

Today's research for the practical application of fusion is carried out mainly on cases of the nuclear reaction of deuterium (D) and tritium (T), i.e., a deuterium-tritium (DT) reaction. Among the above, it has the lowest threshold and is the least difficult to achieve. The advantage of the DT reaction is that it generates more energy than the deuterium-deuterium reaction (DD reaction). However, the DT reaction has a much lower energy threshold than the other listed reactions. The energy threshold of a nuclear fusion reaction is the lowest kinetic energy that the nuclei involved in the reaction must contain in order to break up the electromagnetic repulsive forces (a detailed insight reveals that this is not an ordinary process but a quantum mechanical tunnel process).

An estimate of the number of fusions of two atomic nuclei per unit of volume over a given time is possible if the rate of thermal motion of these two nuclei is known, i.e., the plasma temperature. The number of fusion reactions per unit of time and volume is called reactivity. Reactivity is obtained by multiplying the cross section for the reaction, which is a function of the relative velocities of D and T, by the relative velocities of these two particles, and by calculating the average according to the velocity distribution of both types of particles:

$$R = \iint \sigma(v')v' f_1(v) f_2(v) d^3 v_1 d^3 v_2, \tag{6.6}$$

where σ is the cross section for the reaction, $v' = v_1 - v_2$ is the average velocity of the particles, v_1 is the velocity of the first particle, v_2 is the velocity of the second particle, and v is the relative velocity. The function of the relative velocities of D and T is as follows:

$$f_j(v_j) = n_j \left(\frac{m_j}{2\pi kT} \right) \exp \left(- \frac{m_j v_j^2}{2kT} \right), \tag{6.7}$$

where m_j is the mass of the nucleus, T is the temperature, and

$$v_j = \sqrt{v_{xj}^2 + v_{yj}^2 + v_{zj}^2}. \tag{6.8}$$

If we introduce the average velocity v and reduce the mass μ and insert the Maxwell distribution function into the above integral, we obtain the following:

$$\begin{aligned} R = n_1 n_2 \frac{(m_1 m_2)^{3/2}}{(2\pi kT)^3} \iint \exp\left(-\frac{m_1 + m_2}{2kT}\left(v + \frac{m_1 - m_2}{2(m_1 + m_2)} v'\right)^2\right) * \\ {}^*\sigma(v')v' \exp\left(-\frac{\mu v' v'}{2kT}\right) d^3v' d^3v, \end{aligned} \tag{6.9}$$

where the average velocity is v and the reduced mass is μ:

$$v = \frac{v_1 + v_2}{2} \text{ and} \tag{6.10}$$

$$\mu = \frac{m_1 m_2}{m_1 + m_2}. \tag{6.11}$$

Integration according to v results in the following:

$$\left(\frac{2\pi kT}{m_1 + m_2}\right)^{3/2}. \tag{6.12}$$

Integration according to v' is integration across the whole velocity, and no longer just across components, so we can write the following:

$$d^3v' = 4\pi v' v' dv', \tag{6.13}$$

And we get the following:

$$R = 4\pi n_1 n_2 \left(\frac{\mu}{2\pi kT}\right)^{3/2} \int_0^\infty \sigma(v') v' v' v' \exp\left(-\frac{\mu v' v'}{2kT}\right) dv'. \tag{6.14}$$

The cross sections for reactions are measured as functions of kinetic energy of the projectile ε but not as functions of the relative velocity of the target and the projectile v'. We introduce the formula:

$$\varepsilon = \frac{m_1 v' v'}{2}. \tag{6.15}$$

According to Eq. 6.14, we can write down the following:

$$R = n_1 n_2 \sqrt{\frac{8}{\pi}} \left(\frac{\mu}{kt}\right)^{3/2} \frac{1}{m_1^2} \int_0^{\infty} \sigma(\varepsilon) \varepsilon \exp\left(-\frac{\mu\varepsilon}{m_1 kT}\right) d\varepsilon n_1 n_2 \langle \sigma v \rangle. \tag{6.16}$$

As mentioned, the DT reaction has the lowest threshold and is the easiest to achieve. For the DT reaction, the maximum value of the integral $<\sigma v>$ is equal to 10^{-15} cm^3s^{-1} or 10^{21} m^3s^{-1} at an average temperature of 55 keV or 6.5*10^8 K. If the densities of deuterium and tritium were equal and if they were of the same size (10^{20} m^3), the reactivity would be around 10^{19} fusions in 1 cubic meter of plasma, in 1 s.

The kinetic energy of nuclei that can be used in a nuclear reaction can be increased by heating. Using Maxwell-Boltzmann's law, we can easily prove that the kinetic energy of a heated gas mixture corresponding to their velocity is proportional to the absolute temperature of the mixture:

$$E = kT \tag{6.17}$$

where k is Boltzman's constant = 8617 × 10^{-5} eV/K.

Kinetic energy of at least 10 keV is required for the DT reaction. From the calculation according to Eq. 6.16, the required temperature for the reaction is as follows:

$$T = \frac{10^4 eV}{8,617 \times 10^5 \frac{eV}{K}} = 1,16 \times 10^8 K \approx 100000000 K. \tag{6.18}$$

A simple calculation shows that a temperature of at least 100 million K is required to start the DT fusion reaction.

To achieve higher impact averages or even a DD fusion reaction, we would need several times higher kinetic energies, from about 40 to 60 keV. The required temperature rises proportionately in these reactions. Due to this, today's fusion research is focused on the DT fusion reaction.

At such extremely high temperatures, the gas mixture is in a complete ionization state. Atomic nuclei are not able to maintain their atomic shells under these conditions, so positive ions (nuclei) and negative ions are mixed as individual particles. This is referred to as high-temperature plasma. The probability of a reaction between plasma particles depends on the plasma density (the number of particles in a given volume). The density can be controlled with the help of external pressure on the plasma.

It follows from the *ideal gas law* that at a given volume the gas pressure is proportional to temperature:

$$pV = \frac{m}{M} RT, \tag{6.19}$$

where the state of the gas is determined by the basic thermodynamic variables pressure *p*, volume *V*, and temperature *T*. With the ideal gas law, one of them can be represented by the other two. $R = 8313$ J/K is the ideal gas constant, *m* is the mass of the gas, and *M* is the mass of kilomole.

Due to the extremely high temperatures, the plasma pressure can only be controlled if the material in the fusion chamber is very low density and if the pressure is very low before the plasma heating begins.

Heated plasma has the property of rapid expansion and thus reduction of density. The number of nuclear fusion reactions in a given plasma mass, at temperature *T*, is a function of the plasma density and the time of the nuclear fusion reaction.

Plasma retention time τ is defined by the ratio of energy and power produced within the time frame of the fusion process. The operation of all previous experimental fusion reactors does not take place continuously (as in fission reactors), but as a series of sequential pulses, each lasting time τ. In future fusion projects, scientists are increasingly shifting plasma time constraints.

An indicator of the efficiency of plasma restriction is the product of plasma density *N* and energy retention time τ. The *Nt* product, better known as the Lawson criterion, shows how a particular fusion system is able to compensate for the energy invested in the system with the energy produced. An analysis of this shows that for successful energy production, the Lawson value is about 5×10^{20} sm^3.

Consider the example of a DT reaction. From a DT reaction, we obtain 17.6 MeV of energy; 14.1 MeV of this energy is absorbed by the neutron produced by fusion. The remaining energy of 3.5 MeV is absorbed by the α particle, the nucleus of the helium atom. The alpha particle has a short range, so it transfers its energy to the plasma, and the neutrons then transfer the energy from the plasma. In addition to the neutron, the energy accumulated in the plasma is also dissipated by radiation, generated by the interaction of neutrons and particles in the plasma. Due to radiation, it is very important that the chamber is clean, without the presence of heavy nuclei. The radiation increases with the successive number of the nuclei.

By releasing energy from the plasma, it cools down. However, with a decrease in plasma temperature, the possibility of a fusion reaction also decreases. Maintaining a balance between the discharged and supplied energy of the plasma is a condition for a self-sustaining fusion process.

By fusing nuclei, we would be able to obtain energy – that is, if we could create more complex conditions, such as sufficient plasma temperature and adequate plasma density. If these conditions were met, the number of fusion reactions per unit of time and volume should be such that the energy released would be greater than the energy required to heat and maintain a suitable plasma.

5.2.1 Plasma Pressure Confinement

The process of fusion in nature takes place inside the stars. The plasma pressure in these systems is naturally regulated and maintained by the gravitational force. In the hydrogen bomb, we establish the necessary pressure conditions and the necessary temperature by explosion. In artificial fusion reaction systems for obtaining fusion energy, we establish an appropriate pressure ratio with the help of two technologies:

1. By means of magnetic confinement fusion (e.g., tokamak)[12]
2. By means of inertial confinement fusion

5.2.2 Magnetic Confinement Fusion Systems

To this day, efforts to obtain more energy than invested have largely been based on the principle of magnetic confinement fusion. Technological plasma confinement is mostly based on the toroidal tokamak[13] solution.[14] The principle of operation of tokamak is quite simple, as shown in Fig. 6.8.

The base of the device, including the tokamak, resembles a transformer. It consists of a circular toroidal chamber that surrounds and retains plasma. The primary thread, located in the axis of the toroidal chamber, creates a magnetic field, the change of which is the appearance of a circular flow of electricity in the chamber, similar to the transformer – the secondary part. The current in the system[15] heats the plasma and creates a magnetic field that pushes the plasma. The second toroidal system creates a magnetic field in the plasma in the direction of flow. The

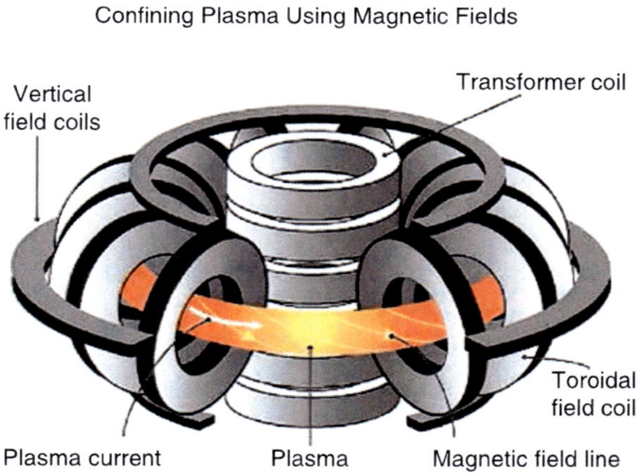

Fig. 6.8 Tokamak – principle of operation

[12] https://sl.wikipedia.org/wiki/Tokamak

[13] A tokamak is a device which uses a powerful magnetic field to confine plasma in the shape of a torus. The tokamak is one of several types of magnetic confinement devices being developed to produce controlled thermonuclear fusion power. As of 2021, it is the leading candidate for a practical fusion reactor. Tokamak was initially conceptualized in the 1950s by Soviet physicists Igor Tamm and Andrei Sakharov.

[14] https://www.iter.org/mach/Tokamak

[15] The current infusion system is several millions of amperes.

field directs the movement of charged particles in the direction of the field forces and thus stabilizes the plasma flow and pushes the plasma against the axis of the torus. The formation of space into a toroidal shape can be explained by the tendency to form a magnetic field and field forces. The result of the combination of both fields produces a stable spiral-shaped electric current of plasma in the toroidal chamber.

The dependence of the plasma pressure and the influence of the magnetic field on the plasma due to compression is marked by β. The factor β determines the degree of plasma compression before the onset of fusion. The higher the factor β, the greater the compression. Figure 6.8 shows that there is another secondary poloid thread in the tokamak device. This serves to shape the plasma and to further stabilize it.

External electricity heats the plasma but not enough to create the necessary conditions for fusion. The magnetic field additionally heats the plasma by means of compression, by bombarding the plasma with neutral atoms of energy 30–100 keV or by high-frequency heating.[16]

Due to the very high temperature of the plasma, it must not come into contact with the internal wall of the toroidal chamber. Any contact of the plasma with the wall of the chamber would lead to a decrease in the plasma temperature and to the contamination of the plasma with impurities. When the plasma is heated, the plasma density is low due to the pressure limitation. During the operation of the fusion reactor, the plasma reaches a density of about 10^{20} ions/m^3. The density of air molecules at normal temperature and pressure is about 10^{25} molecules/m^3. The structure of plasma and air density is of course not comparable, but a comparison of the number of particles in the plasma chamber and in the air indicates an extremely low particle density.

In some of the existing systems, even in some of the best tokamak devices, the plasma maintenance time lasts from 30 s to a few minutes. Future projects using fusion for energy production are expected to reach values from 0.06 s up to 5 s and achieve a maximum excess energy with respect to the energy input.

5.2.3 Technological Issues in Building Tokamak Fusion Reactors

Research in the field of fusion energy has shown that there are certain implementation and technological issues associated with the construction of a fusion reactor. One of the basic problems is related to the characteristics of the materials from which the basic equipment is built and the technological procedures for the use of these materials in the propulsion and support systems of the fusion device. All these materials must meet specific conditions in terms of temperature, the effects of neutron radiation, and deterioration.

One of the most important challenges is thus related to the radiation and temperature suitability and durability of materials in the fusion chamber, especially those

[16] In high-frequency heating, frequencies comparable to the resonance of particles in plasma are used.

that make up the internal walls and plates for controlling the plasma flow. The walls are made of stainless steel with a protective layer made from elements with a low atomic number (carbon, beryllium), to which a part of heavy metals (tungsten, vanadium, niobium) has been added to improve the material. The internal chamber is subjected to collisions of fast neutrons with 14 MeV energy, to X-ray, electron-, and heavy charged particles' radiation, and to the thermal effects of radiation. The change in the structure of the materials of the first wall of the toroidal chamber, due to the action of fast neutrons, requires its frequent replacement, unless an appropriate protection solution is found.

Neutron radiation also affects magnetic compositions, which is reflected in the heating of materials, structural changes in magnetic materials, and changes in insulating materials. It is therefore necessary to improve the technological process of manufacturing magnets, in order to adapt them to the conditions required by such devices.

The preparation of lithium for tritium production is another issue. Lithium is highly reactive with chemical reactions that have a large number of particles. It is also flammable in air, similar to sodium in fast reactors. Problems arise in inventing the optimal structure of the lithium casing (hard, liquid), and in addition, there are difficulties with the cooling process (water, liquid metal).

5.2.4 Safety Aspects of a Nuclear Fusion Reactor

It is generally considered that a fusion nuclear reactor is safer than a fission nuclear reactor. The advantages of a fusion reactor from a safety point of view are mainly the following:

- There can be no uncontrolled increase in power in a fusion reactor.
- There is less radioactive material in a fusion reactor.
- The fusion reactor has no fission products and less radioactive waste.
- When a fusion reactor is shut down, the emission of heat is negligible.
- In the event of an accident, contamination of the surroundings with radioactive radiation is negligible.

On the other hand, some of the basic security concerns in a nuclear fusion reactor are associated with the following:

- Tritium retention in the fusion process.
- Activation of the material of the toroidal chamber in the neutron field and the need for the replacement of the toroidal chamber. This represents great stress for the environment, because the material it is made from is highly radioactive. In the project ITER, the total mass of waste will be over 20 million kilograms, of which 70% will be low-level radioactive waste. Around ten million kilograms of radioactive waste is also expected to be produced during the entire operating time of the project, as a result of replacing equipment and material components.

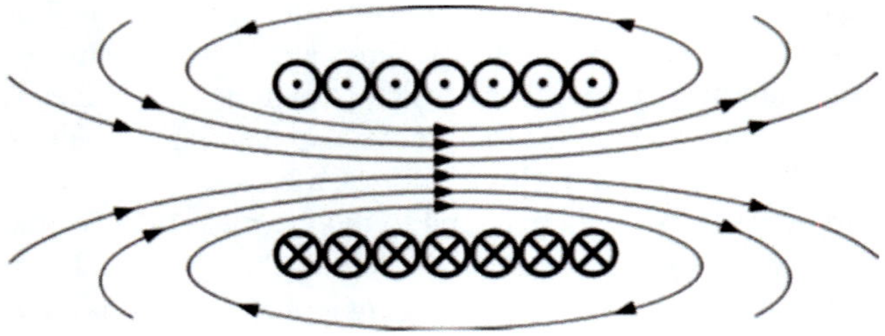

Fig. 6.9 A solenoid magnetic field (coil)

Around 100,000 kilograms will also be represented by plasma flow direction plates, which represent medium or high radioactive waste.
- The very powerful magnetic fields of a tokamak device are a danger to living organisms.

Possible failure scenarios that need to be analyzed:

- Consequences of the diversion of the energy retained in the plasma.
- The consequences caused by the great magnetic energy in superconducting magnets.
- Lithium fires.

The first two failure scenarios involve certain limitations in the operation of the reactor. A possible greater difficulty is lithium fires, which may lead to the leakage of tritium and other activation products into the environment. However, today, the level of risk for the wider surroundings of a fusion reactor is estimated to be lower in comparison to the risks associated with fission reactors.

5.2.5 Magnetic Mirror Plasma Confinement

The Lawrence Livermore National Laboratory, based in the United States, conducts research in the field of fusion energy production in systems with flat solenoid magnetic fields and tandem magnetic mirrors at the ends of the solenoid (the tandem mirror approach) (Fig. 6.9).[17]

From a specific point of view, such a system has its advantages in comparison to tokamak systems, namely:

- Better magnetic plasma confinement
- Higher plasma density
- The possibility of building a system with less power
- Smaller losses at the ends of the coil

[17] A three-dimensional coil that creates a magnetic field (Fig. 6.20). Solenoids are important for their ability to create a controlled magnetic field, so they are very useful as electromagnets.

However, most of today's fusion energy production projects are based on magnetic plasma confinement using large tokamak systems. We will therefore not go into the details of such experimental concepts here.

6 Fusion Research of Inertial Plasma

Fusion devices based on *inertial plasma confinement* operate differently than devices with magnetic plasma confinement. The basic principle of operation is an extremely high deuterium and tritium pressure. The density of deuterium and tritium increases up to a thousand times during the process. Pulses last approximately 10^{-11} s during the inertial plasma confinement procedure. This ensures a suitable value of the Lawson criterion, around 10^{20} sm^{-3}, which is similar to the value in tokamak systems. In inertial plasma confinement systems, the plasma density is much higher than in tokamak systems, so the time limit of the plasma is much lower.

The technique of *compressing plasma* to such extremely high values was developed in 1972 in the United States by experts from the Lawrence Livermore National Laboratory. In this fusion technique, the fuel is stored in miniature pellets containing about 1 mg of a frozen mixture of deuterium and tritium and a casing made of a suitable polymeric material or beryllium. This kind of fuel is then bombarded with laser pulses of extremely high power. A huge amount of energy is transferred to the casings of the pellets, due to which the outer part expands, and the inner part is compressed toward the center of the pellet at a high speed, around 3×10^5 m/s. The core of the pellet, which contains the fusion fuel, is thus compressed with great force. At the same time, the pressure increases by an average of 10^8 bar. Throughout the process, it is important that no additional heating occurs, as heating inhibits fuel concentration. When the maximum compression of the substance is reached, the temperature of the DT mass increases to about 10^8 K. This meets the conditions for a fusion reaction, i.e., a miniature thermonuclear explosion. The result is the expansion of matter and the emission of a neutron pulse. The energy produced by this small explosion is 300 MJ in an extremely short time, of the order of 10^{-10} s. With existing projects, it is possible to achieve the repetition of short intervals with a frequency of 10 times per second.

The key points of inertial plasma confinement systems are the high intensity of the laser beam and the appropriate design of the fuel pellets. The pellets are distinguished by their design and size, depending on the power of the laser beam. Thus, for larger laser beams, pellets with a diameter of up to 5 mm are used.

An efficient fusion system operating on the principle of inertial plasma confinement requires special, very powerful lasers. Such high-power lasers have been developed by the US military, so the specific structure and composition of certain types of lasers are not generally known or available to the wider public.

The basic radiation systems which are used to transform the fuel pellets are as follows:

1. Laser beam systems
2. Light ion beam systems
3. Heavy ion beam system

The first radiation system used in inertial confinement fusion research, and also the most explored, is the *laser beam system.* The light from a laser, which may be of different duration and shape and of a selected wavelength, is concentrated onto a small volume, surrounded and protected by a capsule. The low energy efficiency prevents this technique from being used in inertial fusion reactors at least until the optical charging efficiency is improved. The world's most advanced laser using such technology is located at the Lawrence Livermore National Laboratory, USA. This fusion exploration tool, also called NOVA, produces 40 kJ of energy with a wavelength of 351×10^{-9} m over a period of 3×10^{-9} s to 4×10^{-9} s.

Another fairly well-researched technology exploits a *system of light ion beams.* In an electrostatic accelerator, the high-energy electrical state is shortened to an interval of 10^{-8} s to 10^{-9} s. The diode is connected to a voltage of a few tens of MV and a current of a few MA, while the anode then emits ions (H^+, Li^+, etc.). The efficiency of these systems ranges from 20% to 25%. Pulses with a potential of 30 MV, an energy of 1 MJ, and a surface density of 1 TW/cm^2 are used in laboratories and devices such as the PBF accelerator (Albuquerque, USA) and the KALIF accelerator (Karlsruhe, Germany).

Scientists are also exploring fusion possibilities using *a system of heavy ion beams.* The results of research conducted on particle accelerators to this day have confirmed the possibility of using this kind of technology for energy production. The high efficiency of these complex accelerators and the extended operating periods point to the advantages of this technology. Because the pulse intensity (several kA) is generated for a shorter time (a few 10 ns), the beam between the source and the target must be compressed (reduced magnitude by rank 5). For this reason, more detailed theoretical and experimental studies investigate the very limited space of the beam charge. An experimental device, GSI (Darmstadt, Germany), which has a maximum energy of 20 MeV/nucleon, is able to analyze various particles, including the heaviest natural element, uranium. With the help of the GSI accelerator, six new chemical elements were discovered, as well as certain medical techniques for the treatment of tumors.

Fusion devices based on inertial plasma confinement which are used in energetics are not sufficiently developed in comparison to magnetic plasma confinement systems. Several project studies are exploring the possible construction of nuclear reactors using this kind of technology, but both systems operate on the same basis of energy transmission and supporting technologies. The fusion chamber in inertial plasma confinement systems contains a layer of lithium in which reactions (Li, n) take place. The transfer, through heat, to a usable form of energy, is based on a similar principle.

Taking into account the fact that research on inertial plasma confinement systems is not happening at the level of international cooperation, it is difficult to predict the possibilities for everyday use in energy production (compared to tokamak-type fusion devices). The reason behind the lack of commitment of many countries to take part in this project is, above all, the possible use of this kind of technology for military purposes.

Be that as it may, what can be concluded based on this research is that all these devices operate with a very low efficiency, and this further reduces their applicability for civilian and commercial use.

It thus follows from all existing research on fusion reactors that only technologies using magnetic plasma confinement provide real possibilities for commercial (civilian) development in the relatively near future, roughly by 2050. In the following section, the focus will be primarily on magnetic plasma confinement system research and the results it yielded.

6.1 Fusion Research in the Field of Magnetic Plasma Confinement

Energy systems of the future are going to have to supply energy in a safe and at the same time ecologically sound way. Future-oriented energy research is carried out in six related projects, which indicate potential results over different periods of time. Fusion energy research has been going on for many years. The financial burden of this research is enormous, but fusion does offer the greatest possibility to be used for the world's future energy supply. Fusion energy is a safe source of energy with unlimited fuel sources, and it is environment-friendly, as no environmentally harmful emissions are generated during its processes. Thus, with fusion nuclear reactors, we obtain energy in a way that is nonpolluting and sustainable.

In Europe, all research on nuclear fusion and the development of magnetic plasma confinement systems in the past has been carried out under the supervision of the European Commission. Due to the good organization and progress of unique development programs, it was possible, in Europe, to construct and use the world's most famous experimental nuclear fusion unit, the JET.[18] At the same time, various nuclear fusion devices have also been developed around the world, as shown in Table 6.3.

For the day-to-day, civilian, and commercial applications of nuclear technology, it will be necessary to test the scientific and technological capacities of nuclear fusion exploitation. European and other research programs in this field were the starting

[18] JET (Joint European Torus) is a project operated by Culham Centre for Fusion Energy, UK, which conducts nuclear fusion experiments up to 16 MW. The project involves all European organizations involved in nuclear fusion research, in collaboration with scientists from around the world.

Table 6.3 The largest tokamak systems in the world and their parameters

Device	Country	Interior radius *a* (m)	Exterior radius *R* (m)	Plasma flow *I* (MA)	Toroidal field *B* (T)	Input power *P* (MW)	Start of operation
ITER	Several countries	2.0	6.2	15	5.3	73+	
JET	EU	1.00	2.96	7.0	3.5	42	1983
JT-60 U	Japan	0.85	3.2	4.5	4.4	40	1991
TFTR	USA	0.85	2.50	2.7	5.6	40	1982 closed
TORE-SUPRA	France	0.80	2.4	2.0	4.2	22	1988
T-15	Russia	0.70	2.4	2.0	4.0	–	1989
DIII-D	USA	0.67	1.67	3.0	2.1	22	1986
ASDEX-U	Germany	0.5	1.67	1.4	3.5	16	1991
TEXTOR-94	Germany	0.46	1.75	0.8	2.6	8	1994
FT-U	Italy	0.31	0.92	1.2	7.5	–	1988
TCV	Switzerland	0.24	0.875	1.2	1.43	4.5	1992
C-MOD	USA	0.22	0.67	1.5	8.07	4.5	1992
MAST	UK	0.5	0.7	2	0.63	6.5	1999
NSTX	USA	0.67	0.85	1.0	0.6	11.5	1999

point for the international project ITER,[19] which demonstrates the technology operating inside a nuclear reactor as an integrated system. More than 2000 scientists are involved in this project, exploring the technological aspects of high-temperature plasma and superconductors. The project also reflects a high level of international industrial cooperation.

7 ITER: International Thermonuclear Experimental Reactor

Research on how to accommodate for the world's future energy supply has been going on for some time. As mentioned earlier, the fusion of light atomic nuclei is the most promising technology within the field of nuclear energy. Under the watchful eye of scientists from around the world, explorations in this field have made great strides in recent decades. The Fusion Development Community is ready for the next step toward a complete energy supply for the future, the ITER project. The purpose of this project, of enormous proportions, is to validate the possibility of using nuclear

[19] ITER (International Thermonuclear Experimental Reactor) is an international project, as part of which an experimental fusion reactor is currently under construction in Cadarache, southern France.

fusion to generate electricity and to obtain important data for the design and management of the first fusion nuclear reactor. ITER and DEMO[20] focus especially on safety aspects and environmental policy, as well as on the socioeconomic level of fusion. They also aim to foster confidence in the general public, regarding nuclear fusion energy.

In order to organize the engineering work as part of the ITER project, several technology sections have been established, with experts working in different fields:

- Superconductivity
- High-temperature flow
- Remote management and control
- Waste management
- Other areas (supply systems, design support, etc.)

Both past and present collaboration with the JET project offer unique knowledge and application of scientific foundations in the field of plasma physics. The results that JET has and continues to offer are crucial in building ITER. They help to address technical problems that arise in the implementation of the project.

The ITER project was founded in 1985 under the guidance of the former Soviet Union, the United States, the European Union, and Japan under the auspices of the IAEA. It marked the next step in the development of magnetic fusion and was considered the successor to the largest tokamak experiment to date. Conceptual planning began in 1988, but the year 1990 is considered a milestone in construction planning. The overall programmatic objective of ITER is "to demonstrate the scientific and technological feasibility of fusion energy for peaceful purposes." ITER offers additional knowledge for designing and demonstrating the operation of a fusion reactor, capable of generating huge amounts of energy. The whole process of obtaining energy must, of course, comply with environmental regulations.

Between 1998 and 2001, the ITER project employed around 500 scientists. The representation of Europeans in the ITER project is currently around 40%. With the help of the ITER project, scientists will explore plasma in conditions similar to those expected in future commercial fusion reactors, which are estimated to produce approximately 500 MW of electricity over a longer period of time. The energy produced will be about ten times higher than the energy needed to maintain the plasma at the required temperature. The ITER project will be the first to generate electricity and test-distribute that energy through a network. It involves testing of some of the key technologies that are essential for the construction of a fusion reactor in the future, such as heating, control, diagnostics, and remote control.

The construction cost of the ITER project is estimated at around € 5 billion in the first 10 years and € 5 billion in the second period, for the next 20 years of operation.

[20]The demonstration power plant DEMO is a project parallel to ITER, which tests the suitability of fuel and the resistance of materials to radiation.

In June 2005, the participating countries designated the town Cadarache in southern France as the site for the construction of ITER. One year later, the ITER organization was founded, which will eventually own the ITER device and which has the task of overseeing all aspects of the project. The planning of the ITER project finished in 2001. The detailed plan for the construction, subsequent operation, and decommissioning was divided into four phases:

- Construction preparation phase
- Construction phase
- Operational phase
- Decommissioning phase

Currently, the project is in the third (operational) phase, which is roughly divided into five time periods. The final operational phase of decommissioning is foreseen within 11 years after the completion of the operational phase. It includes device deactivation, disassembly, and removal of all components.

7.1 *Technical Objectives of ITER*

The ITER project aims to achieve greater energy production than the system requires to operate. The factor Q shows the amount of thermal energy produced in the fusion reactor divided by the energy supplied. If the value of Q is less than 1, it means that the energy produced is less than the energy required for heating the plasma and thus for the operation of the system. The largest tokamak JET test device to date had the factor $Q = 0.65$. With ITER it is supposed to increase to a value of $Q = 10$ (or to a value of $Q > 5$ in a steady-state operation). The device should demonstrate the availability and integration of technologies essential for a fusion reactor (such as superconducting magnets and remote maintenance) and test tritium breeding module concepts that would lead in a future reactor to tritium self-sufficiency, the extraction of high-grade heat and electricity (Table 6.4).

Table 6.4 Parameters of the ITER

Total fusion power	500 MW (700 MW)
Q = fusion power/external input power	>10
Average cross section of the "neutron wall"	0.57 MW/m^2
Time of inductive burning of the plasma	>300 s
Exterior plasma diameter	6.2 m
Interior plasma diameter	2.0 m
Toroidal field (diameter 6.2 m)	5.3 T
Plasma volume	837 m^3
Plasma surface	678 m^2

7.1.1 Model Reactor and Costs

Some facts about fusion reactors and fusion thermonuclear power plants:

- Reactors exploiting the DT reaction and cooling with water or helium will use steam and gas turbines.
- Fusion power plants will have a similar capacity to today's coal-fired power plants and nuclear power plants, i.e., in the range 500–1500 MW.
- The price of electricity from future fusion power plants will be approximately the same as the price of electricity from today's thermal power plants, including the cost of the entire service life (decommissioning and storage of radioactive waste).
- The costs of the ITER project are not comparable with future fusion power plants, as there will be differences in the materials used; the serial production of such power plants will be cheaper, the physical size of power plants will be different, etc.

In light of the above facts, the following equation shows that the price of the entire power plant will deviate only slightly from the cost of the reactor:

$$P_{\text{gen}} = P_{\text{fus}}(0,2 + 0,8\,M)\eta - P_{\text{internal}} - P_{\text{fus}}/Q/h \tag{6.20}$$

where:

- P_{gen} transported power into the network
- P_{fus} fusion power
- M enhancement of neutron power
- η thermal conversion efficiency
- P_{internal} internal use of electrical power (without heating)
- Q increase in plasma power
- h electrical energy conversion efficiency (for heating)

Results of calculations show that the costs of reactors are comparable to the costs of ITER; therefore, there is a good chance that the price of electricity generated in the future will be comparable to today's electricity prices from thermal power plants and today's nuclear power plants, regardless of the fact that the costs of current power plants do not include environmental costs, which will be significantly lower in the case of fusion power plants.

The last 50 years of fusion research and development have brought ever new challenges for scientists and engineers of at least two generations. It is not yet possible today to anticipate certain problems that will arise in the course of developments. Therefore, ITER is crucial for the development of new technologies. When asked when fusion would be ready for use, Leo Artsimovitch, the father of the tokamak device, once replied: "Fusion will be useful, when humanity will need it."

The development of fusion represents several unique aspects:

- The development of fusion has specific goals, which require a longer period of time in order to accommodate for one of basic human needs – the need for energy.

- It requires multigenerational connections between scientific knowledge and knowledge of the humanities and engineering.
- It requires a multidisciplinary approach to problem solving.
- Because research in this field is "open-ended," it requires the participation of scientists from different perspectives and political backgrounds to openly share their knowledge.

7.1.2 The Safety Aspect

The extraction and exploitation of fusion energy is known as a relatively safe and clean source of energy, but a cautious approach is still needed to achieve these goals. It should be noted that fusion reactors also have an impact on the natural environment. ITER affects the environment in the following ways:

- Various risks arising in the construction phase, in the operation phase, and in the decommissioning phase
- Accidents that may occur
- Waste products

The greatest risks are tritium, as a fuel, and radioactive dust, which is caused by the erosion of components when they come into contact with the plasma. Tritium has a radioactive half-life of 12.5 years. In the event that the environment is contaminated with tritium, the biological half-life that affects humans is much shorter, from 10 to 15 days. In fusion power plants, almost all of the tritium produced will be consumed, except for a small negligible part. Throughout the life span of the ITER system, 16 kg of externally supplied tritium is expected to be consumed. The existing production and global tritium stocks are limited due to very little consumption. Because of the prescribed limitations regarding the volume of tritium that is allowed to be transported, the ITER system requires weekly transportation of tritium. For safety reasons, the maximum amount of tritium allowed in the area where ITER operates is limited to 3 kg. Thus, in the event of an accident, the maximum total quantity involved in the accident is limited to 0.450 kg.

The construction of the reactor and the support systems is designed in such a way so as to minimize the risk of failure. The multilevel approach to building the system reduces the possibility of environmental pollution. To avoid possible accidents, detritification air-water systems and filtration systems have been set up, which ensure a minimum pollution of the environment. The impact of radioactive radiation on the most at-risk groups of persons during the normal operation of ITER is less than 1% of the natural background radiation.[21]

[21] Radiation that originates from natural sources and to which people are exposed in everyday, ordinary life, e.g., the radiation of rocks and the Earth as opposed to radiation caused by man.

Based on analyses, some of the most (unlikely) scenarios of possible accidents involving tritium, active corrosion of materials, products, and the cooling system of neutron-activated tokamak powder on the environment are anticipated:

- Abnormal operation of plasma physics
- Reduction or incorrect cooling flow in the primary system
- Interruption of cooling of the primary system
- Interruption or malfunctions of external cooling systems
- FT coil terminal faults, etc.

Possible errors can add certain additional radioactive doses to natural background doses, which is a negligible contribution to the natural background. Security analyses do not show a major impact or risks of environmental contamination. ITER's ultimate safety measures address the analysis and hypothetical events of potential accidents. Even under the blackest fault scenario, the design concepts and operational characteristics of the reactor are designed so that there is no need to evacuate the surrounding residents.

Possible failures can add additional radioactive doses to the dose of natural background radiation, but these would be minimal. Safety analyses do not show a major risk of environmental contamination. ITER's safety measures are based on analyses and a careful consideration of hypothetical events in the event of potential accidents. Even with a worst-case scenario, the design concepts and operational characteristics of the reactor are designed so that there is no need for evacuation of the surrounding public.

During the entire lifespan of the ITER project, a considerable amount of radioactive material will be generated. Most radioactive material will be recycled over a period of 50–100 years. Radioactive material that will be generated during the entire operating life of the power plant and after decommissioning is divided into the following:

- Radioactive material resulting from neutron fusion
- Contaminated materials (beryllium, tungsten in the form of tokamak powder)
- Active corrosion products
- Tritium
- Products of the above

High-level radioactive material, whose radioactivity does not decrease in 100 years, is classified as radioactive waste that requires a special repository under appropriate conditions for a longer storage period. After the decommissioning of the facilities, 30,000 tons of radioactive material are expected, 80% of which will be decontaminated over a period of 100 years. After 100 years, ITER waste will pose a much lower risk to the environment than waste produced by fission nuclear power plants or even the larger coal-fired power plants of today. In comparison to a fission power plant, a fusion reactor produces 50,000 times less pollution.

The ITER project also includes a special security training program for staff employed during the project construction time, the operational phase, and the decommissioning phase. It combines radiation protection as well as protection against conventional risks. The purpose of the security program is to protect personnel, device components, and maintenance procedures.

7.1.3 A Vision of Fusion Energy: A Continuation of the ITER Project

Fusion is the nuclear technology which is to provide the world with sustainable energy supply (Parisi & Ball, 2019). Much scientific progress has been made in this field. Today the major obstacles on the way to nuclear fusion power plants are no longer related to principal scientific questions but rather to questions of design and suitable technical solutions. European fusion research is focused on the ITER project.

Preparation for the construction of a commercial fusion reactor is planned to be finished in 2050. Even though this sounds far ahead in the future, it is actually a very optimistic time schedule. During the past 50 years of fusion research, the difficulties of developing a fusion power plant have been underestimated, and timelines for realization have been postponed again and again.

The prediction in 2007 was that the feasibility of constructing and operating fusion plants will depend on their safety level, environmental impact, and economic viability. These factors are hard to predict because the outcome depends on a range of assumptions regarding future power plant designs and future structural materials. Surprising developments cannot be excluded during the research process. In particular the technical realization of a material which provides a shield against neutron radiation and is not quickly degrading due to activation and at the same time is a big challenge. This could even require new test facilities before an applicable solution is found. Fusion power plants will not be ready in time to contribute timely to the CO_2 reductions. Even when finished, they will be of limited use because of the very high investment costs.

But let us start from the beginning and consider where we are today, 15 years after this pessimistic prognosis.

7.1.4 The Situation Today: 2020

ITER's First Plasma is scheduled for December 2025. That will be the first time the machine is powered on and the first act of ITER's multi-decade operational program. On a cleared, 42-hectare site in the south of France, building has been underway since 2010. The central tokamak building was handed over to the ITER organization in March 2020 for the start of machine assembly. The first major event of this new phase was the installation of the 1250 ton cryostat base in May 2020. In the ITER offices around the world, the exact sequence of assembly events has been carefully orchestrated and coordinated.

The successful integration and assembly of over one million components (ten million parts), built in the ITER members' factories around the world and delivered to the ITER site, constitute a tremendous logistics and engineering challenge. The

ITER organization will be carrying out the work supported by a number of assembly contractors. In November 2017, the project passed the halfway mark to First Plasma. In July 2020, the project officially launched the machine assembly phase. Today, project execution to First Plasma stands at 71.1 percent (October 2020 data).

ITER is designed to operate with a nominal fusion power of 500 MW. A project of approximately equal dimensions and costs, DEMO, is envisaged as a continuation of the ITER project. The DEMO fusion reactor will be the first project to generate electricity, thus making the first commercial use of fusion energy possible. The fusion power should be increased by a factor of 4 to bring the potential of electric power supplied to the electricity grid to about 500 MW, which is comparable to one of today's smaller power plants. The heat of such a system will be four times higher than in the ITER project, which consequently requires about four times the plasma efficiency. Meeting these conditions will require an approximately 15% increase in the linear dimension system and an approximately 30% increase in the density of magnetic confinement plasma. The exact requirements will be indicated by the results of ITER.

The timeline shows that the use of the first commercial fusion power plant of the tokamak type before 2050 will be impossible. In case of successful completion of the ITER project, the DEMO project will serve as the first prototype of the commercial use of fusion energy. The projected progress in the development of fusion energy is shown in Fig. 6.10. The PROTO fusion reactor is envisaged as the first commercial version of a fusion power plant to serve as a prototype for further mass exploitation of fusion energy, with a doubling of electrical power, at the expense of only 10% of increase in the dimensions of the system.

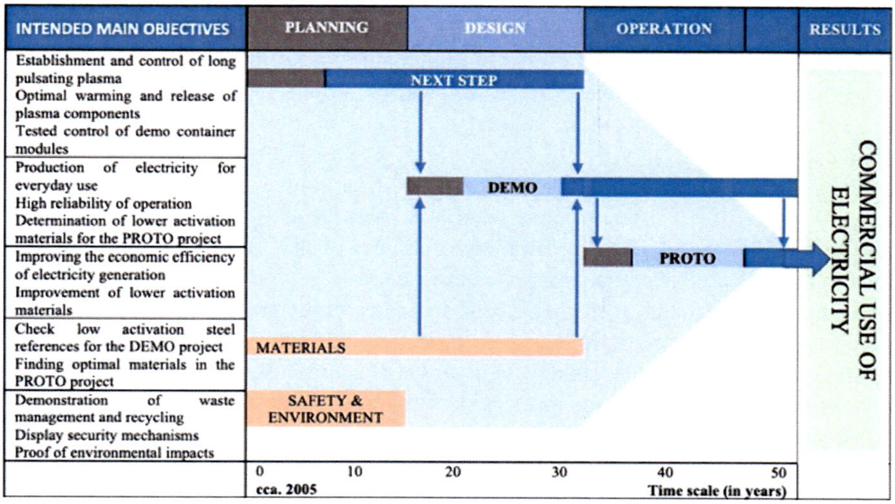

Fig. 6.10 Projected development of fusion energy until commercial use

The economic viability of fusion reactors combines solutions from several areas. The expected total costs of the power plant, calculated according to the volume of tokamak, are as follows:

- DEMO project – 14 €/W
- PROTO project – 8 €/W
- serial production PROTO – reduced cost – 4 €/W

Such values are comparable to the costs of thermal power plants and fission nuclear power plants. These range between 1.5 and 3 €/W. However, these costs for coal-fired power plants do not take into account the costs of environmental pollution and do not include fuel costs, operating costs, and decommissioning costs. These also reach comparable values, around 3 €/W, so today's cost is somewhere between 4.5 and 6 €/W.

7.2 *Other Low-Carbon Nonsustainable Solutions*

7.2.1 Natural Gas vs. Low Carbon

In the mid-2010s, it became common to say that natural gas would be a bridge fuel to a zero-carbon future, in which solar, wind, and other renewable technologies provide all our energy without any carbon dioxide emissions to worsen climate change. But if natural gas is really a bridge, then it's not part of the long-term plan. And if we actually build the bridge, we're likely to stay on it. Natural gas consumption in the United States has risen by a third in the past 15 years. Gas accounts for 32% of total energy consumption and is now the biggest source of electricity nationwide, largely displacing coal-fired power plants. Natural gas, primarily methane, burns much cleaner than coal does, and it provides ready backup to variable wind and solar farms. That sounds promising, except burning natural gas still creates CO_2. Methane in wells and pipelines can leak into the atmosphere, amplifying global warming. And once the last coal plant closes, natural gas plants become the dirtiest electricity sources (Weber, 2021).

To reduce CO_2 emissions, society has to decarbonize its energy systems as quickly as possible. Building more wind and solar farms is relatively inexpensive and fast, and it accelerates the shutdown of coal plants. But exploiting the best locations, the windswept plains and sunny deserts, requires a greatly expanded transmission grid to bring the electrons to major cities and manufacturing complexes. Those wires and poles introduce risks from windstorms, floods, and fires, all rising because of climate change, and township after township routinely fights expansion plans: "Not in my backyard." And more electricity does not readily solve the need for liquid fuels burned in trucks, ships, and planes or for intense heat in industrial foundries, distilleries, and refineries that make volumes of metals, cement, glass, jet fuel, and chemicals. The energy density of liquid fuels is difficult to match.

If we can clean emissions out of the natural gas system, it could be part of a carbon-neutral future instead of a bridge. The technology exists to extract the carbon or to transform the gas so that carbon coming out and carbon going in balance to zero or near zero. The first step in a comprehensive plan for decarbonizing the nation's energy infrastructure would be improving energy efficiency and conservation to reduce consumption (McKinnon, 2018a, b). The second would be to electrify as many cars, space heaters, water heaters, and stoves as is practical, using renewable sources. At the same time, tighten up the leaky gas infrastructure. And replace as much natural gas as possible with low-carbon alternatives such as biogas, hydrogen, and synthesized methane, or use a process called pyrolysis at the end of the natural gas pipes to get the carbon out. Clean energy supporters rightly worry that any investment in gas infrastructure creates a lock-in effect. Each new power plant, pipeline, or gas storage unit has an expected lifetime of 25–80 years, so each element could either become a trap for more emissions or a stranded asset.

But we can solve the lock-in problem with drop-in alternatives to natural gas: *low-carbon gases* that can flow through existing pipes, tanks, and power plants, taking advantage of those trillions of dollars of assets.

7.3 *Low-Carbon Gas vs. Zero-Carbon Gas*

The drop-in substitute most ready for natural gas is biomethane (methane gas produced from biological sources). Microbes inside large drums called anaerobic digesters chew up organic matter such as crop waste, manure, sewage, and food waste and other garbage in landfills, producing methane. Biodigesters, already a mature technology, transform waste streams at landfills and the waste lagoons adjacent to concentrated animal feeding operations from environmental liabilities into valuable commodities, generating revenues for municipalities and farmers.

Biogas can serve as a direct substitute for natural gas, but the relative volume, globally, is low. If a farm, landfill, or sewage plant cannot readily use the gas to make electricity or is not next to a gas grid, the biomethane might need to be liquefied and trucked to another location, reducing the carbon payoff. Still, biomethane is a commercially ready technology that can begin to decarbonize part of the gas system (McKinnon, 2018a, b) (Fig. 6.11).

7.4 *Hydrogen Instead of Methane*

Natural gas can be replaced entirely with hydrogen. Turbines can burn hydrogen to generate electricity for the grid, and internal combustion engines can burn it in heavy-duty vehicles. Hydrogen in fuel cells can produce electricity for commercial use. And hydrogen is a ready building block for many basic chemicals. Burning it, or

Strategies for Decarbonizing the Gas System: Replacing Gas with Biomethane

To reduce CO_2, emissions from burning natural gas, which is primary methane, the gas in a pipeline can be replaced with biomethane. Microbes inside large drums break down organic matter, crop waste, food waste, etc. to produce methane, water vapor and some. CO_2 the last of which is captured for industry or underground storage. If the matter had decomposed naturally, it would have released CO_2 and methane into the atmosphere.

Add bacteria to organic matter to create CO_2 water and methane

Fig. 6.11 Strategies for decarbonizing the gas system: replacing gas with biomethane

reacting it in fuel cells, does not produce CO_2. Leaked hydrogen has a warming effect much lower than that of methane.

In order to exploit solar energy to a full extent, we need to find a suitable way of transforming it into energy that can accumulate in large quantities. This could be created by hydrogen production. Hydrogen can be produced by electrolysis of water in special solar power plants. Such power plants would be located on uninhabited areas, with intense solar radiation (deserts). The hydrogen thus produced could be stored in a gaseous state underground, e.g., in depleted natural gas fields, or in the liquefied state in special tanks. It would be transported by special ships or by gas pipelines.

Hydrogen is an ideal fuel for heating, as only water vapor is produced during combustion, which does not require chimneys. However, it can also be used as a fuel for fuel cells, which are used to produce electricity with what is known as the cooling process. Of course, hydrogen can also be used as fuel in steam boilers in thermal power plants to produce electricity. It should be emphasized that hydrogen is an extremely clean fuel which virtually does not pollute the environment at all.

Natural hydrogen seeps out of the ground from basins in large blocks of rock that form parts of continents, and scientists have discovered this more than 100 years ago. Oil and gas companies, however, whenever hydrogen was found next to underground reservoirs, have considered it a problem because it catches fire easily and corrodes metal piping. Today, researchers are drilling hydrogen test wells and launching long-term programs to search for hydrogen underground. An important resource is out there; we only need to figure out how to exploit it cheaply and safely.

Hydrogen can be produced artificially. Currently, most hydrogen for industrial use is produced from steam-methane reforming, i.e., adding heat and hot water to methane to create hydrogen and CO_2. Hydrogen gas is also a product of electrolysis (the process of using electricity to split water into hydrogen and oxygen). Both these processes require significant amounts of energy, however (Fig. 6.12).

Moving and storing gaseous hydrogen are also a challenge. Because of hydrogen's low density, it takes a lot of energy to move it through a pipe compared with denser gases such as methane or liquids such as petroleum. After several hundred

Strategies for Decarbonizing the Gas System: Replacing Gas with Hydrogene

Natural gas in a pipeline can be partly or fully replaced with hydrogen, which can be burned in power plants and heavy-duty vehicles, fed into fuel cells for car, homes and buildings, or used as an industrial feedstock. Burning hydrogen emits no carbon. Special pipelines alloys would be needed to transport gas with greater than 20% hydrogen. Several sources of hydrogen are described below.

1. NATURAL RESERVOIRS

When water underground contacts iron-bearing rock, it turns the rock into silicates and hydrogen. Or pumps can send water down to convert the rock

Water	naturally reacts with	rocks containing iron	to create	silicates	and	Hydrogen

2. STEAM RE-FORMING OF METHANE, WITH CARBONE CAPTURE

Heat and water added to methane from a well created hydrogen and CO_2 which would capture rather then emitted. Industry already uses this process widely to make hydrogen for refineries and foundries. Clean energy would be required to generate the heat.

Add	heat	to	CH_4	and	steam (H_2O)	to create	CO_2	and	H_2

3. ELECROLYSES

Renewable electricity splits water to create hydrogen. More energy is needed than for steam re-forming of methane, but the only by-product is oxygen, which is released into the air.

Add	electricity	to	water	to create	oxygen	and	hydrogen

Fig. 6.12 Strategies for decarbonizing the gas system: replacing gas with hydrogen

kilometers, the inefficiency makes moving hydrogen more expensive than the value of the energy it carries. And hydrogen can embrittle steel pipelines unless that is mitigated by altering operating conditions or incorporating expensive alloys (McKinnon et al., 2013).

One way to integrate hydrogen is to mix it with methane in an existing natural gas pipeline. Such blending decarbonizes some of the system by displacing a portion of the natural gas with hydrogen. Experiments in the United Kingdom and France have shown that a mixture of 80% methane and 20% hydrogen can be efficiently moved in a natural gas pipeline.

Fittings inside furnaces and stoves, such as burner tips, might need to be altered or replaced for blends with more than 20% hydrogen because, like pure hydrogen, blended gas burns at different temperatures and rates. Another consideration is that because of hydrogen's low energy density, a 20% blend by volume provides 14% less energy per cubic foot than natural gas.

One way around certain cost and safety challenges is to pipe hydrogen as part of another chemical form we know how to handle, such as ammonia, which has one nitrogen atom and three hydrogen atoms. Molecules that include hydrogen atoms are known as hydrogen carriers. Hydrogen is converted, where it is found or produced, into the carrier, which is dropped into existing pipelines, and is either used in that form or converted again into hydrogen at the destination. Common carriers such as ammonia, formic acid, and methanol are liquid at near-ambient conditions, making them easier to transport than gaseous hydrogen. Although ammonia is caustic, it is already moved worldwide as a fertilizer ingredient, and it can be burned without producing any CO_2. Methane could be the most efficient option because it carries four hydrogen atoms for every carbon atom and is already compatible with existing pipes, compressors, tanks, turbines, and appliances.

Demonstration projects are growing quickly in number. Since 2003, Wärtsilä and Eidesvik Offshore have collaborated on the development and verification of new environmental technology.[22] Many in the shipping industry consider ammonia to be one of the future fuel options for achieving zero emissions. Together with ShipFC and partners Eidesvik Offshore, Equinor, and Prototech, Wärtsilä will now determine if this is the case. In a 5-year demo project, the involved parties will test a new energy system based on ammonia-powered fuel cells. The solution will be tested on Eidesvik's supply vessel, and the technology will make the 17-year-old workhorse the world's first zero emissions supply vessel when green ammonia is used as fuel. Finnish industrial builder Wärtsilä is constructing a new ship for 2023 named Viking Energy that will run on ammonia with fuel cells, avoiding greenhouse gas emissions and other pollutants that plague the maritime sector. "Fuel cell modules with a combined effect of 2 MW will be installed on board the Viking Energy. The vessel will operate at sea for one year with the new energy system, and the goal is to be able to document zero emissions from the vessel during the period it is fueled by ammonia. For the shipping industry, such results will mean that we are one step closer to the IMO's goal of a minimum 50 per cent reduction in greenhouse gas emissions from international shipping traffic by 2050," says Ingve Sørfonn, head of technology and development in Marine Electrical Systems at Wärtsilä.

Air France and the Charles De Gaulle airport in Paris are very interested in hydrogen as a way to decarbonize aviation. Hydrogen carriers are still in the early stages of research, however, so it is difficult to say how successful they could be.

Power plants that burn hydrogen are being designed as well. In Delta, Utah, the Intermountain Power Plant, one of the largest US coal-fired plants, sends electricity hundreds of miles to Los Angeles. To meet the city's long-term requirement for renewable and low-carbon energy, in 2025, plant owners will replace the coal boilers with turbines that can burn hydrogen. They will start with a blend of 30% hydrogen in natural gas and will shift to 100% hydrogen later. The hydrogen will be generated right there using electrolysis powered by wind and solar and will be stored in more than 100 existing, underground salt caverns, each about the size of the Empire State Building.

[22] marit.holmlund-sund@wartsila.com

7.4.1 Fuel Cells

For more than 150 years, scientists have known that when hydrogen and oxygen combine to form water (H_2O), the chemical reaction releases energy. Devices that use a controlled combination of the two gases to generate energy in the form of electrical current are called fuel cells. This developing technology underlies the vision of a "hydrogen economy," in which the only exhaust from fuel cell-powered vehicles would be water vapor and the world would drastically reduce its dependence on petroleum-based fuels.

The Physical Background

Fuel cells are electrochemical devices that serve to directly convert chemical energy into electricity. Fuel cells, just like batteries, consist of two electrodes immersed in the same electrolyte. The fuel (a chemical element or compound that has a high internal energy content) at the anode of the fuel cell oxidizes, and the electrons released during fuel oxidation are led from the anode with the outer circuit of the conductor, through the consumer (resistor, DC electric motor, light bulbs, etc.), to the cathode. At the cathode, another element (an oxidant) is reduced by trapping electrons produced at the anode. The results of the reaction, the negative and positive ions, combine in the electrolyte, and the resulting product is discharged from the fuel cell. Often, the final product of the reaction is the same as that of a fuel burned in the oxidant upon direct conversion of the chemical into internal thermal energy. Hence the name, fuel cell.

Fuel cells are high-efficiency energy converters. They are without moving parts, so they work without noise. Their use is currently limited to spacecraft and certain military uses.

According to the mode of operation, fuel cells are classified into the following:

- *Primary*, where fuel and oxidant are conducted from external tanks and the reaction results in ejection. Such an element is shown in Fig. 6.13a.
- *Secondary*, regenerative fuel cells, where the product of the reaction is restored to the original elements, as shown in Fig. 6.13b, with a supply of energy (e.g., thermal or electrical energy). Solar energy, or energy from nuclear reactors, can

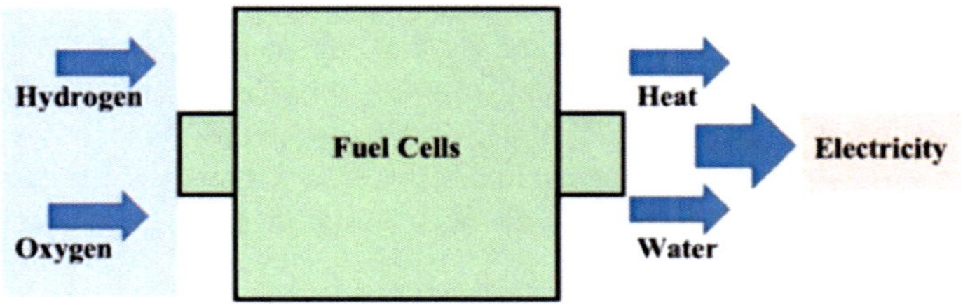

Fig. 6.13a Primary fuel cells

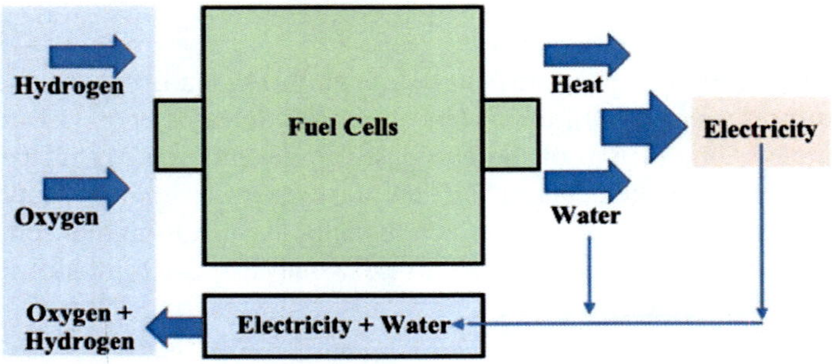

Fig. 6.13b Secondary fuel cells

also be used for regeneration. Elements can be renewed inside or outside the fuel cell, continuously or in cycles.

In principle, primary fuel cells function in a similar way as batteries. Secondary fuel cells operate in a similar manner as accumulators. Both types of fuel cells are simply a means of indirect energy storage.

Historical Overview

The fuel cell was discovered in 1839 by W.R. Grove, who observed the reverse electrolysis process. He found that if oxygen was supplied to one electrode and hydrogen to the other, electricity was produced. He described his experiments in 1842, referring to the fuel cell as a *gas voltaic battery*.

Grove was the first to observe that an electric current is produced at the point of contact of three phases:

- Gaseous (hydrogen or oxygen)
- Liquid (electrolyte)
- Solid (platinum electrodes)

He also found that the voltage depends on the *active surface* of the electrode. With this observation, he not only discovered the fuel cell but also set the course for the development of this field which would follow in the next century.

Despite the fact that the work of many well-known chemists and electrochemists focused on the development, theory, and experimentation with fuel cells, technically acceptable solutions were only proposed by the English researcher F.T. Bacon in 1932. In the mid-1950s, the first fuel cells suitable for vehicle propulsion and small electrical appliances were produced. A significant contribution to their development was made by space research programs. In recent decades, however, the acute energy crisis and environmental pollution have been a significant impetus for fuel cell research.

7.4.2 The Principle of Operation of a Fuel Cell

The simplest way to explain the principle of operation of a fuel cell is by referring to the most highly developed fuel cell system so far, i.e., using hydrogen as a fuel and oxygen as an oxidant. If hydrogen and oxygen (both in a gaseous state) are brought to contact, they react (they combine into water), thereby releasing energy.

$$2\,H_2 + O_2 \rightarrow 2\,H_2O + \text{ENERGY} \tag{6.21a}$$

In a fuel cell, this joint reaction consists of two reactions, each of which takes place at one electrode. Hydrogen is oxidized at the anode, releasing electrons.

$$H_2 \rightarrow 2H^+ + 2e \tag{6.21b}$$

Electrons are transferred through an external circuit to the cathode, where oxygen is reduced:

$$O_2 + 2H_2O + 4e \rightarrow 4\,OH^- \tag{6.22}$$

Oxygen reduction is a complex, multistage reaction whose mechanism depends on the nature of the electrode surface. The whole reaction consists of several sub-reactions (sub-steps)

$$O_2 \rightarrow O + O\ (\text{dissociation}) \tag{6.23}$$

in which atomic oxygen is adsorbed on the surface of the metal electrode. Through a reaction with water:

$$O + H_2O \rightarrow 2OH\ (\text{hydroxylation}) \tag{6.24}$$

hydroxyl groups are formed, adsorbed as hydroxide on the surface of the metal, and finally reduced to ions:

$$OH + e \rightarrow OH^-\,(\text{electronation, reduction}) \tag{6.25}$$

In an acid electrolyte fuel cell, hydrogen ions generated at the anode travel through the electrolyte and fuse in the reaction layer of the cathode with hydroxyl ions into water, as shown in Fig. 6.14. The electrolyte in the fuel cells may also be alkaline. In this case, the hydroxyl ions enter the reaction layer of the anode through the electrolyte by diffusion; there they recombine into water with hydrogen ions. There are many other fuel cell mechanisms in which an oxidant is reduced and various molecules are oxidized, which serve as fuel. The above example demonstrates the complexity of electrochemical reactions and the concerns faced by fuel cell designers.

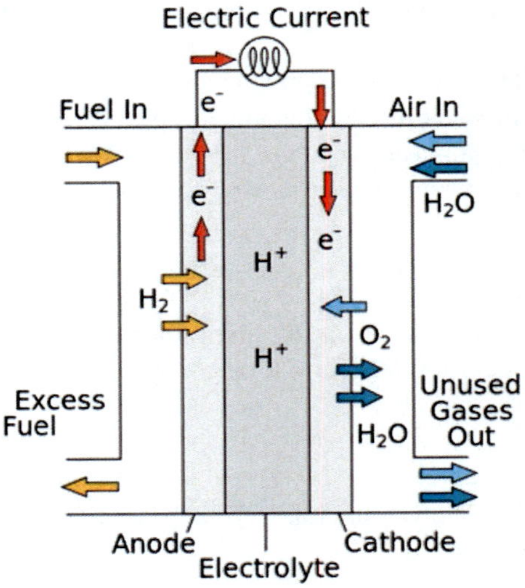

Fig. 6.14 Schematic diagram of a fuel cell with hydrogen, oxygen, and an acidic electrolyte

7.4.3 Fuel Cell Elements

Fuel cell elements include the following:

- Electrodes
- Catalysts
- Electrolytes
- Fuels and oxidants

Electrodes

Electrochemical reactions of oxidation and reduction take place at the electrodes of the fuel cell. The electrodes have a multiple function. Because they translate electrons, they are made of metals or materials with semiconductor properties. Electrodes should have appropriate mechanical properties, a large surface area, and a suitable pore size. The electrodes must not corrode in the electrolyte of the fuel cell. There are only a few metals that meet these requirements, such as platinum, palladium, radium, ruthenium, and, to some extent, nickel for anodes and silver for cathodes. Graphite is also a good material for the electrodes of fuel cells operating at medium and high temperatures. Because precious metals are rare and expensive, newer electrodes often use them only in the form of fine dispersions in the matrix of a polymer (Teflon, polyethylene), on a carrier made of nickel, copper, or some other common metal. The widespread use of fuel cells in the future depends mainly on the discovery of other, cheaper materials for electrodes. The most important property of electrodes is their porosity. The gas entering the pores of the electrode must come into contact with the part moistened by the electrolyte. At that point, the gas ionizes,

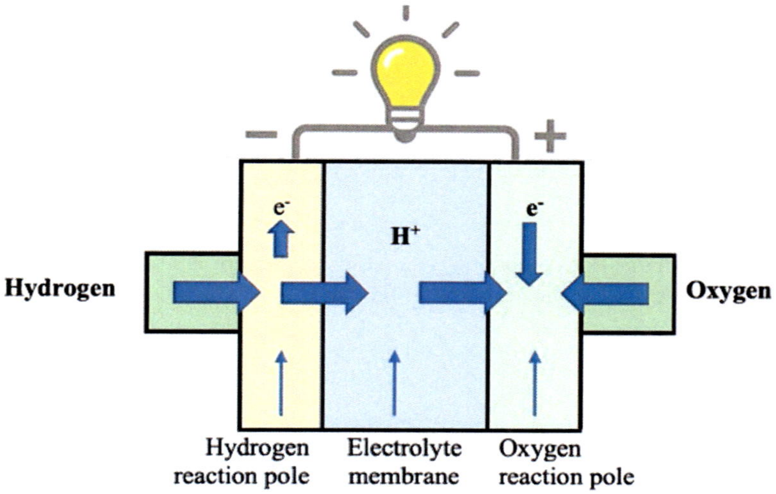

Fig. 6.15 Schematic of an electrode

while the electrolyte receives and discharges the generated ions. The calculation of the pore diameter, the inlet gas pressure, and the state of the electrolyte surface are critical elements of the fuel cell electrode technology. Electrodes with pores of different sizes proved to be the best. In such electrodes, the pores on the electrolyte side are narrow, so they have a great capillary effect. On the gas side, they are wide so that gas can easily enter (Figs. 6.14 and 6.15). In this way, we can regulate the reaction area.

Catalysts

The current of the electrochemical reaction and the current density depend on the catalytic phenomena on the electrode surface. The electrode, too, can serve as a catalyst (i.e., electrocatalyst) if it is treated by a special process or if the catalyst is adsorbed onto its surface. The choice of catalyst depends on the slowest elemental stage of the whole reaction, which needs to be either accelerated or reduced.

Electrolytes

In an electrolyte, the charges are transferred by means of diffusion of ions from one electrode to another. We distinguish between solid and liquid electrolytes. An electrolyte can be any polar liquid that dissolves ionic crystals, molten salts, and most often alkali metal salts (chlorides and carbonates). The most important solid electrolytes are membrane-built polymers (e.g., polystyrene). The membranes separate the cathode from the anode part and reduce the dimensions of the fuel cell.

Such electrolytes are mainly used in spacecraft because the operation of a fuel cell with such an electrolyte does not depend on gravitational forces.

Recently, electrolytically conductive membranes made of nickel boride or boron nitride have been investigated.

Fuels

Any chemical element or compound with a high content of internal energy that can release this energy by oxidation and turn into ions can act as fuel in a fuel cell.

Oxidants

An oxidant is any chemical substance which has a strong affinity for electrons and which, by reduction, turns into an ionic state. The most common oxidant is oxygen. In some reactions, halogen elements (Cl, Br, F) are also used as oxidants.

7.4.4 Fuel Cell Efficiency

The efficiency of fuel cells can be defined in several ways: as thermodynamic, voltage, or faradaic efficiency.

Thermodynamic Efficiency

The maximum possible energy that can be obtained from a chemical reaction, the so-called free energy of a chemical reaction ΔG, is defined as follows:

$$\Delta G = \Delta H - T\Delta S \tag{6.26}$$

where ΔH is the total free energy (reaction enthalpy). T is the reaction temperature and ΔS is the change in entropy. The following equation applies to electrochemical energy converters (batteries, fuel cells):

$$\Delta G = -n\,F\,E \tag{6.27}$$

where n is the number of equivalents (equal to the number of electrons), F is the amount of electricity bound to one equivalent (Faraday constant), and E is the electromotive force of the element.

Table 6.5 Thermodynamic efficiency

Code	Carriers of electric	Transmission direction	Efficiency
Alkaline FC	OH^-	Cathode–anode	60%
FC with solid polymer	H^+	Anode–cathode	45–60%
Phosphor acid FC	H^+	Anode–cathode	40%
FC with melted carbonate	$CO_3{}^{2-}$	Cathode–anode	65%
FC with solid oxide	O^{2-}	Cathode–anode	50%

Thus, thermodynamic efficiency is defined by the equation:

$$\eta_i = \frac{\Delta G}{\Delta H} = 1 - T\frac{\Delta S}{\Delta H} = 1 - n\,F\,T\frac{\left(\frac{dE}{dT}\right)_p}{\Delta H} \tag{6.28}$$

where at a constant pressure p:

$$\Delta S = -\left(\frac{\delta}{\delta T}\Delta G\right) \tag{6.29}$$

Thermodynamic efficiency for various reactions is listed in Table 6.5.

It can be seen from the table that the thermodynamic efficiency can also be greater than 1 if the change in entropy, i.e., the coefficient of change of electromotive force with temperature, is positive.

$$\eta_v = \frac{V}{E} \tag{6.30}$$

Voltage Efficiency

Due to internal resistances (electrolyte resistance, electrode reaction layer resistance, etc.), an internal voltage drop occurs so that the voltage at the contacts is lesser than the electromotive force E. Based on this, the voltage efficiency of the fuel cell is defined.

Faradaic Efficiency

It is a measure of the completeness of fuel utilization in an electrochemical reaction that produces electricity. It is defined as follows:

$$\eta_m = \frac{Q}{Q_m} \tag{6.31}$$

where Q is the amount of electricity (C, As) obtained from the fuel element and Q_m is the amount of electricity that could be obtained from the fuel in a hypothetical (ideal) electrochemical reaction. Q is always less than Q_m.

The total efficiency of the fuel cell is the product of all three:

$$\eta_T = \eta_i \, \eta_v \, \eta_F \tag{6.32}$$

The overall efficiency is lower than the thermodynamic efficiency mainly due to the internal voltage drop and unwanted changes that occur in the fuel cell during operation. The efficiency achieved so far for hydrogen-oxygen-type fuel cells is about 60%.

7.4.5 Types of Fuel Cells and Potential Use

Fuel cells are made to order. The only series-standardized types are the low-temperature fuel cells of the American company General Electric, which were produced for the Gemini (1 kW) and Apollo (1.5 kW) space programs. All other models are still under research. For some of them, the basic information is given in Table 6.6.

The justification for the future use of fuel cells is based on two factors:

- *High level of efficiency*, much higher than in any known thermal process so far.
- *Low negative impact on the environment*, as the combustion products do not contain harmful ingredients. In addition, they operate silently.

Table 6.6 Characteristics of fuel cells (FC)

Code	Abbrev.	Working temperature	Electrolyte	Fuel	Oxidant	Use
Alkaline FC	AFC	<120 °C	Potassium hydroxide (KOH	Hydrogen	Oxygen	Space technology
FC with solid polymer	SPFC	100 °C	Solid polymer	Hydrogen methanol	Oxygen, air	Vehicle, low power plants
Phosphoric acid FC	PAFC	200 °C	Phosphoric acid	Natural gas, methanol	Air	Heat plants
FC with melted carbonate	MCFC	650 °C	Lithium and potassium carbonate	Natural gas	Air	Power plants, heat plants
FC with solid oxide	SOFC	1000 °C	Zirconium oxide (ZrO_2)	Natural gas, alcohols	Air	Power plants, heat plants

7.4.6 Research and Applications

Significant research on fuel cells has been conducted to accelerate their development and successful introduction into the marketplace. Because fuel cells allow the regulated reaction of hydrogen (in a tank) and oxygen (from the air) to produce electricity, they are often perceived as nonpolluting. However, that describes only the exhaust from the vehicle, not the processes involved in producing hydrogen gas. Hydrogen (like electricity) is not a primary source of energy but rather an energy carrier. There are no natural reservoirs of pure hydrogen. To produce hydrogen, water can be electrolyzed in the reverse of the fuel cell reaction. To force electrolysis to occur, considerable amounts of electricity are required, much of which is currently generated by burning fossil fuels. Alternatively, in commercial applications, the vast preponderance of hydrogen generated in the world is extracted from a fossil fuel: natural gas. When natural gas (basically methane, a lightweight molecule made of carbon and hydrogen) is exposed to steam under high temperatures in the presence of a catalyst, it frees the hydrogen. This is called "reforming," and the process produces carbon dioxide (CO_2). Discussions about increasing hydrogen production by gas reforming are often accompanied by calls for carbon sequestration schemes.

Not all hydrogen fuel cells, however, are destined for vehicles. Stationary fuel cells for electric power generation have been under development for decades. Some applications of fuel cells to residential or commercial buildings could involve generating electricity from a fuel input like natural gas or hydrogen and using the waste heat from that process to heat the building.

Still, the costs of hydrogen energy remain high compared to other energy sources; it is extremely difficult to store onboard, the use of natural gas as a feedstock emits CO_2, and there are numerous technical challenges to overcome with these systems. It could take many years of research and development, as well as changes in the energy infrastructure, before hydrogen could have a significant impact on world energy use.

Nonetheless, the use of hydrogen for vehicle fuel has the potential to nearly eliminate gasoline use in light-duty vehicles by the middle of this century, while reducing fleet greenhouse gas emissions to less than 20% of current levels.

Before fuel cells can be used universally, two key issues will need to be addressed:

- Substituting precious metals (from which electrodes are made) with a different, cheaper material
- Addressing the durability and reliability of fuel cells

If these two fundamental setbacks are successfully resolved, fuel cells will most likely be used to propel road vehicles. This means that the engine would be a DC electric motor. Fuel and oxidant would be pumped into the tanks of vehicles at gas stations similar to the ones we use today. The use of hydrogen fuel cells as fuel for stable high-level power plants is also being investigated. Hydrogen would be produced by electrolysis of water and with unused energy from thermal, wind, or solar power plants when the load on the electrical system would be low. This would increase the efficiency of power plants and, above all, enable the storage of energy in the form of hydrogen.

7.5 End of the Pipe

Instead of decarbonizing natural gas before it goes into the pipeline, we could remove the carbon at the end of the pipe, where customers consume the gas. Methane, for example, can be split at the user's location into hydrogen and solid carbon, which looks like a fine, black dust. The process, called methane pyrolysis, is efficient and eliminates CO_2 emissions. Every kilogram of hydrogen produced from pyrolyzed methane generates 3 kilograms of solid carbon instead of 9 kilograms of CO_2 gas that would be emitted if the methane was burned.

The pile of carbon dust that accumulates inside a collector in a furnace or stove would be carted away each month or so. We already pay garbage haulers and municipal wastewater-processing plants to clean up our solid and liquid wastes; we should pay to clean up the waste from our gas use, too. The carbon piles actually have value, though, because they can be sold as a basic ingredient for making graphite, rubber, coatings, batteries, and chemicals, as well as a soil amendment for agriculture.

Although engineers have studied methane pyrolysis for decades, they have deployed it only in small demonstration projects. Some equipment at the end of the pipe has to be changed to separate the carbon, but no expensive hydrogen pipelines would have to be built, simplifying matters greatly. Pyrolysis of conventional natural gas can bring the entire system to nearly zero carbon. Adding methane from biodigesters or made from CO_2 in the atmosphere using renewable electricity could make the system carbon-negative.

Imagining any of these decarbonized futures might conjure up visions of large new industrial complexes or millions of small equipment changes for consumers. But so do other proposals to curb emissions. Electrifying every heater, stove, and vehicle would require widespread technology replacement. Plans to directly pull CO_2 from the air would require millions of big machines to capture the gas and sequester it, thereby sprawling enterprises that would also demand lots of new land and new electricity (Figs. 6.16 and 6.17).

Decarbonized gas would let us take advantage of trillions of dollars of existing pipelines, equipment, and appliances, saving huge sums of money and years of time in creating a zero-carbon energy system. We would, of course, have to fix the leaky infrastructure. Leaks can be minimized by replacing pneumatic equipment with electric devices at well sites, improving the automation of pipe and tank inspections with sensors on drones and robots, and writing regulations that no longer turn a blind eye to leaking, as well as deliberately venting or burning unwanted gas. This work would create jobs for workers in the oil and gas industries and would clean up the energy infrastructure, which in turn could lessen pollution in communities near energy facilities.

Transport Hydrogen in Other Chemical Forms

Hydrogen 'carriers' are gases or liquids that contain hydrogen molecules, such as formic acid (CH_2O_2) methanol (CH_3OH) or methane (CH_4). Instead of extracting natural gas and burning it, thereby creationg CO_2, machines powered by renewable electricity pull CO_2 from the air or from smokestacks, then combine it with hydrogen to from the carier, which is shipped in existing pipelines. Burning it releases CO_2 again, but because the fuels vas made with CO_2 from atmosphere, net emission are low.

Add	CO_2	to	H_2	to create	H_2O	and	CH_4

Fig. 6.16 Transport hydrogen in other chemical form

Extract Carbon at Consumer Location

Carbon can be pulled from natural gas at the consumer end of the pipeline instead of the supplier's end. In a process called pyrolysis, an appliance split the incoming methane into hydrogen gas for heaters, stovers or industrioal machines, leaving solid carbon dust that the appliance collects. Consumer or haulers cart away the dust, which could be sold as a raw material for production such as batteries or fertilizer. Plasma-an electrically charged gas (like a lit neon sign)-or heat splits the methane.

Add plasma or	heat	to	CH_4	to create	solid C	and	H_2

Fig. 6.17 Extract carbon at consumer location

Reining in climate change requires many solutions. Declaring who cannot be part of those, such as natural gas companies, only raises resistance to progress. Because decarbonized gas can complement renewable electricity and because it might be a faster, cheaper, and more effective path for parts of society that are difficult to electrify, we should not discard gas as an option. We have a massive gas infrastructure, and we have to figure out what to do with it. Scrapping it would be slow, expensive, and incredibly difficult, but we could instead put it to work to help create a low-carbon future.

7.6 Photosynthesis

In search of the most ecologically sound technologies, scientists have often imitated natural processes. In this spirit, the idea of mimicking natural photosynthesis was born, the attempt to create an "artificial leaf." In this section, we will focus on what natural photosynthesis is. We know that this seemingly simple process actually incorporates many complex processes, and the simplicity is in fact deceptive. We will consider the process of natural photosynthesis and analyze specific concepts related to it, before turning to the concept of artificial photosynthesis (Bryant & Frigaard, 2006).

1. Natural Photosynthesis

Natural photosynthesis is (in simplified terms) the conversion of light energy into chemical energy (oxidation of water, photolysis), where oxygen is released, carbon dioxide is reduced, and organic compounds (sugars, amino acids, fats) are formed.

The process of photosynthesis takes place in chloroplasts. A simplified formula for this process is as follows:

$$6CO_2 \textit{ (carbon dioxide)} + 6H_2O \textit{ (water)} \xrightarrow[\text{Light}]{} C_6H_{12}O_6 \textit{ (sugar)} + 6O_2 \textit{ (oxygen)}$$

The name photosynthesis comes from the Greek *phōs (φῶς)*, which means "light," and *sunthesis (σύνθεσις)*, which means "putting together."

Photosynthesis is influenced by three main factors, which are as follows:

- Light intensity and wavelength
- Carbon dioxide concentration
- Temperature

In addition, photosynthesis is also limited by a number of other environmental factors, such as the amount of light available, the amount of leaf area (shading by other plants can be a major limitation of photosynthesis), the rate of carbon dioxide supply, water availability, etc., according to Chapin et al. (2002). The rate of photosynthesis depends on various factors, such as the quantity of water, temperature, amount of light, concentration of carbon dioxide, environment pollution, etc.

At the molecular level, the process of photosynthesis is divided into light reactions and dark reactions, also called the Calvin cycle or the carbon cycle. In light reactions, powered by the sunlight, adenosine triphosphate (ATP) as an energy source and nicotinamide adenine dinucleotide phosphate (NADPH) as a reducing agent are formed. In dark reactions, CO_2 is used to produce complex organic compounds, for which ATP and NADPH are essential (Lehninger et al., 2008).

Photoautotrophs are organisms that perform photosynthesis, that is, most plants, most algae, and *cyanobacteria.*[23] These organisms are able to synthesize food from carbon dioxide and water, using light. In *photoheterotrophs*, organic compounds are used as a carbon source instead of carbon dioxide (Bryant & Frigaard, 2006). This process is responsible for producing and maintaining the oxygen content of the Earth's atmosphere, and photosynthesis is also important in providing the energy we need for life on Earth (Bryant & Frigaard, 2006). As mentioned, two kinds of processes occur in natural photosynthesis, light reactions (light-dependent reactions) and carbon fixation (dark reactions).

[23] Cyanobacteria, also called *blue-green algae*, are microscopic organisms found naturally in all types of water. These organisms use sunlight to make their own food.

- *Light stage* (light-dependent reactions):

  This stage involves the conversion of light energy into chemical energy and takes place on the thylakoid membranes of the chloroplast, consuming water and light and producing oxygen, adenosine triphosphate (ATP), and nicotinamide adenine dinucleotide phosphate (NADPH). In this stage, chlorophylls and other auxiliary pigments absorb sunlight and then synthesize ATP and NADPH with the help of sunlight. Oxygen is formed as a by-product. Light-dependent reactions take two forms, cyclic and noncyclic.

  The formula for light-dependent reactions in green plants is as follows (Raven et al., 2005):

$$\begin{aligned} & 2H_2O + 2NADP + +3ADP + 3Pi + light \\ & \quad \rightarrow 2NADPH + 2H + +3ATP + O_2 \end{aligned} \tag{6.33}$$

The flow of electrons from water to NADP+ is explained by the Hill-Bendall Z-scheme.[24]

- *Dark stage* (carbon fixation or carbon-fixing reactions):

  The dark stage involves the reduction of CO_2 to produce sugars, amino acids, fats, and so on. This stage takes place in the stroma (matrix)[25] of the chloroplast; ATP and NADPH are used to reduce CO_2 during carbon fixation. This produces carbohydrates (e.g., triose phosphates, sucrose, starch).

  This part of photosynthesis is also called the Calvin cycle, the stage when carbon dioxide, ATP, and NADH+H+ enter the equation. In dark reactions, the enzyme RuBisCO[26] captures CO_2 from the atmosphere and uses NADPH in the process (Calvin cycle), releasing three-carbon sugars. These sugars later combine to form sucrose and starch. The general equation for light-independent reactions in green plants is as follows (Raven et al., 2005) (Fig. 6.18):

$$\begin{aligned} 3CO_2 + 9ATP + 6\ NADPH + 6H+ \rightarrow\ & C_3H_6O_3\text{-phosphate} + 9ADP \\ & + 8Pi + 6NADP + 3H_2O \end{aligned} \tag{6.34}$$

[24]The Z scheme shows *the pathway of electron transfer from water to NADP+*. Using this pathway, plants transform light energy into electrical energy and, hence, into chemical energy as reduced NADPH and ATP. Hill and Bendall proposed the Z scheme.

[25]Stromal tissue is *primarily made of extracellular matrix containing connective tissue cells*. Extracellular matrix is primarily composed of ground substance – a porous, hydrated gel, made mainly from proteoglycan aggregates – and connective tissue fibers.

[26]*RuBisCo, (rubisco)* is abbreviation for ribulose-1,5-bisphosphate carboxylase-oxygenase and is an enzyme involved in the first major step of carbon fixation, a process by which atmospheric carbon dioxide is converted by plants and other photosynthetic organisms to energy-rich molecules such as glucose.

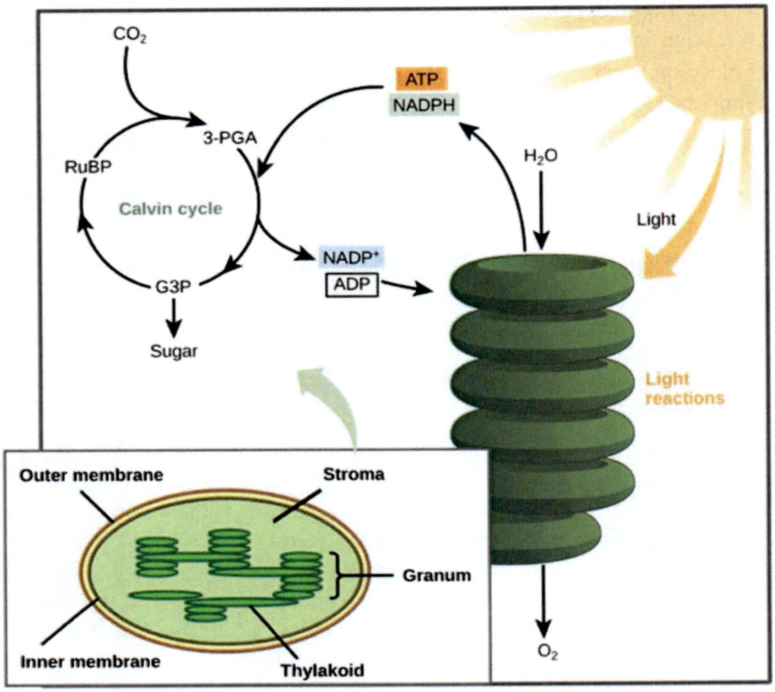

Fig. 6.18 Photosynthesis

Photosynthesis takes place in plants and some bacteria; hence we distinguish between:

- Plant photosynthesis, characteristic of green plants and cyanobacteria
- Bacterial photosynthesis, which occurs in green and purple bacteria

> *In plants, algae and cyanobacteria, long-term energy storage in the form of sugars is produced by a subsequent sequence of light-independent reactions called the Calvin cycle*[27]*; in other bacteria, different mechanisms such as the reverse Krebs cycle are used to achieve the same end*[28] (Online Etymology Dictionary, 2020).[29]

[27]The *Calvin cycle* is the cycle of chemical reactions performed by plants to "fix" carbon from CO_2 into three-carbon sugars. ... The Calvin cycle is also sometimes referred to as the "light-independent" reactions of photosynthesis, since it is not powered directly by photons from the sun.

[28]The *Krebs cycle* is a sequence of reactions in the living organism in which oxidation of acetic acid or acetyl equivalent provides energy for storage in phosphate bonds (as in ATP) – called also citric acid cycle, tricarboxylic acid cycle.

[29]http://654.pl/lp/wiki/view.php?sq=streaming&lang=en&q=Photosynthesis

As Govindjee points out, the efficiency of plant photosynthesis can vary depending on various factors such as the frequency of converted light, light intensity, temperature, proportion of carbon dioxide in the atmosphere, etc. (Govindjee, 2014). Plants usually convert light into chemical energy with a photosynthetic efficiency of 3–6% of total solar radiation (Miyamoto, 1994, 2013). As Maxwell and Johnson state, absorbed light that is not converted is scattered primarily as heat. Quite a small proportion and about 1–2% is reemitted as chlorophyll fluorescence at longer (red) wavelengths (Maxwell & Johnson, 2000).

2. Artificial Photosynthesis

Artificial photosynthesis is a chemical process that mimics the natural process of photosynthesis (Rasul et al., 2017). As mentioned earlier, natural photosynthesis involves the conversion of sunlight, water, and carbon dioxide into carbohydrates, energy, and oxygen. Artificial photosynthesis is a kind of imitation of natural photosynthesis, which takes place under certain specific conditions. The following section provides a comparison between the processes of natural and artificial photosynthesis. In the latter, carbon dioxide (CO_2) could be used as a raw material for the production of fuel for generating environment-friendly energy.

The term artificial photosynthesis is commonly used to refer to any scheme for capturing and storing the energy from sunlight in the chemical bonds of a fuel (a solar fuel), whereby photocatalytic water splitting converts water into hydrogen and oxygen. Artificial photosynthesis is a major research topic, which numerous present-day scientists set out to explore. Below are some of their basic ideas explained in short.

Quader and Ahmed state:

> *Artificial photosynthesis, which is described as 'Chemistry's Greatest Challenge,' attempts to replicate the natural processes of photosynthesis, and, at least in near future, the goal of artificial photosynthesis is to use sunlight energy to make high-energy chemicals to store energy* (Quader & Ahmed, 2017).

Nawaz et al. describe it as follows:

> *Artificial photosynthesis is referred as 'a mimic of photosynthesis process' that harvest natural sunlight and in turn convert the CO_2 into fuels like hydrogen and carbon based. The hydrogen is produced from water splitting while carbon-based fuels like methanol, methane, or CO are usually produced from the reduction of CO_2 in the presence of water* (Nawaz et al., 2019).

7.6.1 Brief Historical Overview

Research on artificial photosynthesis began to blossom at the beginning of the twenty-first century, though a few other similar ideas have been put out there years before that. The very idea of abandoning or reducing the use of fossil fuels dates back around a hundred years. Italian chemist Giacomo Ciamician, also known as the father of solar energy, proposed a transition from the use of fossil fuels to solar energy in one of his lectures, in 1912, saying:

Where vegetation is rich, photochemistry may be left to the plants and by rational cultivation, as I have already explained, solar radiation may be used for industrial purposes. In the desert regions, unadapted to any kind of cultivation, photochemistry will artificially put their solar energy to practical uses. On the arid lands there will spring up industrial colonies without smoke and without smokestacks; forests of glass tubes will extend over the plants and glass buildings will rise everywhere; inside of these will take place the photochemical processes that hitherto have been the guarded secret of the plants, but that will have been mastered by human industry which will know how to make them bear even more abundant fruit than nature, for nature is not in a hurry and mankind is. And if in a distant future the supply of coal becomes completely exhausted, civilization will not be checked by that, for life and civilization will continue as long as the sun shines! If our black and nervous civilization, based on coal, shall be followed by a quieter civilization based on the utilization of solar energy, that will not be harmful to progress and to human happiness (Venturi et al., 2005).

Another name worth mentioning is Akira Fujishima, who "discovered the photocatalytic properties of titanium dioxide, the so-called Honda-Fujishima effect, which could be used for hydrolysis" (Fujishima et al., 2000). As artificial photosynthesis is a central theme in the research sphere of today, it has inspired the formation of a number of consortia for artificial photosynthesis. Among them is the one established by the Uppsala University, Sweden, in 1994, as a collaboration between three universities, in the Swedish towns of Lund, Uppsala, and Stockholm. This consortium conducts basic research "on the common goal of producing hydrogen from renewable solar and water sources" (Swedish Consortium for Artificial Photosynthesis, 2020). Their vision is to develop new ways to produce hydrogen from solar energy and water. The Joint Center for Artificial Photosynthesis (JCAP)[30] has been operating since 2010 and is run by a team from the California Institute of Technology (Caltech):

JCAP was established in 2010 as a U.S. Department of Energy Innovation Hub that aims to find new and effective ways to produce fuels using only sunlight, water, and carbon dioxide (JCAP, 2020).

The field of artificial photosynthesis has continued to receive attention, and there is currently a great deal of research dedicated to achieving a common goal:

This field of research, called artificial photosynthesis, is currently the subject of intense interest, driven by the aim of converting solar energy into the carbon-free fuel hydrogen through the light-driven water splitting (Andreiadis et al., 2011, p. 946).

The difference between natural and artificial photosynthesis is illustrated in Fig. 6.19 (Hong et al., 2014; Khoulya et al., 2017):

As Yarris (2009) argues, the goal of scientists is to develop an artificial version of photosynthesis, which could be used to produce liquid fuels from carbon dioxide and water (Yarris, 2009). Researchers from the US Department of Energy's Lawrence Berkeley National Laboratory (Berkeley Lab) took a critical step toward this, as Yarris argues, by discovering that "cobalt oxide crystals can effectively perform a critical photosynthetic reaction of water molecule cleavage" (Yarris, 2009). Authors Wang et al. say that obtaining solar energy to convert CO_2 into chemical fuels is "a

[30] https://solarfuelshub.org

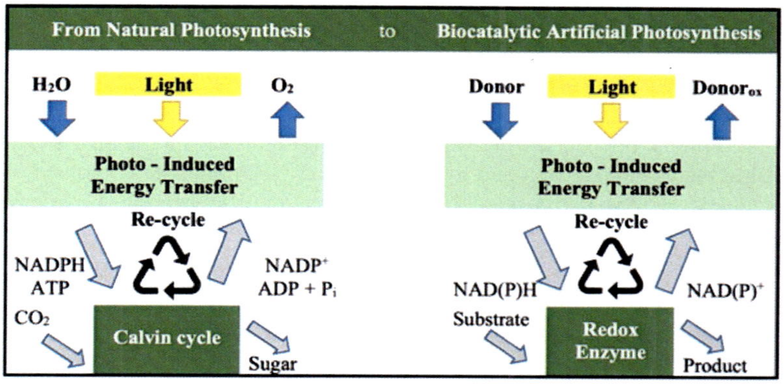

Fig. 6.19 Natural and artificial photosynthesis

promising technology to reduce rising atmospheric CO_2 levels and alleviate global dependence on fossil fuels" (Wang et al., 2020a, b).

Research in this promising field has also been the center of attention for Heinz Frei (Berkeley Lab), and Feng Jiao, who "reported the results of their study in the journal *Angewandte Chemie* in an article entitled: *Nanostructured Cobalt Oxide Clusters and Mesoporous Silica as Efficient Oxygen-Evolving Catalysts* (2009)." This research was conducted at the Helios Solar Energy Research Center (Helios SERC), a scientific program at the Berkeley Lab led by Paul Alivisatos (Jiao & Frei, 2009; Yarris, 2009). As Yarris says, this process has positive implications:

> *Artificial photosynthesis for the production of liquid fuels offers the promise of a renewable and carbon-neutral source of transportation energy, meaning it would not contribute to the global warming that results from the burning of oil and coal. The idea is to improve upon the process that has long-served green plants and certain bacteria by integrating into a single platform light-harvesting system that can capture solar photons and catalytic systems that can oxidize water – in other words, an artificial leaf* (Yarris, 2009).

Frei attempted to explain how the principle of artificial photosynthesis works in the first place:

> *To take advantage of the flexibility and precision by which light absorption, charge transport and catalytic properties can be controlled by discrete inorganic molecular structures, we have been working with polynuclear metal oxide nanoclusters in silica," Frei said. "In earlier work, we found that iridium oxide was efficient and fast enough to do the job, but iridium is the least abundant metal on earth and not suitable for use on a very large scale. We needed a metal that was equally effective but far more abundant* (Jiao & Frei, 2009; Yarris, 2009).

Like any technology, however, this one also has its challenges. Authors like Wang et al. draw attention to "the composition of efficient and robust systems for selective CO_2 photo conversion free of harmful reagents and external biases" (Wang et al., 2020a, b). Their solution is a "photocatalyst sheet," "a potentially extensible CO_2 exploitation technology which converts CO_2 and H_2O to formats and O_2"

(Wang et al., 2020a, b). Some authors, such as Reece et al. contrastingly, describe "the development of solar cells that 'divide' water and contain earth elements that operate in almost neutral pH conditions, both with and without connecting wires" (Reece et al., 2011). As Van Noorden adds, "a 'slice' with a catalyst coating is a silicon version of a photosynthetic sheet: it turns sunlight into fuel, which it can store by splitting water into oxygen and hydrogen" (Van Noorden, 2012). Some authors, such as Calzaferri, are also researching various materials that would be suitable for artificial photosynthesis (Calzaferri, 2010).

As mentioned, Wang et al. postulate that obtaining solar energy to convert CO_2 into chemical fuels is "a promising technology to reduce rising atmospheric CO_2 levels and alleviate global dependence on fossil fuels" (Wang et al., 2020b). Here are some details from their summary of the article (2020b) entitled "Molecularly Engineered Photocatalyst Sheet for Scalable Solar Formate Production from Carbon Dioxide and Water":

> *Here we present a photocatalyst sheet that converts CO_2 and H_2O into formate and O_2 as a potentially scalable technology for CO_2 utilization. This technology integrates lanthanum- and rhodium-doped $SrTiO_3$ ($SrTiO_3$:La,Rh) and molybdenum-doped $BiVO_4$ ($BiVO_4$:Mo) light absorbers modified by phosphonated Co(II) bis (terpyridine) and RuO_2 catalysts onto a gold layer. The monolithic device provides a solar-to-formate conversion efficiency of $0.08 \pm 0.01\%$ with a selectivity for formate of $97 \pm 3\%$. As the device operates wirelessly and uses water as an electron donor, it offers a versatile strategy toward scalable and sustainable CO_2 reduction using molecular-based hybrid photocatalysts* (Wang et al., 2020b, p. 703).

One of the studies by Pang et al. suggests innovative solutions for the development of artificial photosynthesis technology. In an article entitled "Enhanced CO Evolution for Photocatalytic Conversion of CO_2 by H_2O over Ca modified Ga_2O_3 (2020)," the authors state:

> *Artificial photosynthesis is a desirable critical technology for the conversion of CO_2 and H_2O, which are abundant raw materials, into fuels and chemical feedstocks. Similar to plant photosynthesis, artificial photosynthesis can produce CO, CH_3OH, CH_4, and preferably higher hydrocarbons from CO_2 using H_2O as an electron donor and solar light. At present, only insufficient amounts of CO_2-reduction products such as CO, CH_3OH, and CH_4 have been obtained using such a photocatalytic and photoelectrochemical conversion process. Here, we demonstrate that photocatalytic CO_2 conversion with a Ag@Cr-decorated mixture of $CaGa_4O_7$-loaded Ga_2O_3 and the CaO photocatalyst leads to a satisfactory CO formation rate (>835 μmol h−1) and excellent selectivity toward CO evolution (95%), with O_2 as the stoichiometric oxidation product of H_2O. Our photocatalytic system can convert CO_2 gas into CO at >1% CO_2 conversion (>11531 ppm CO) at ambient temperatures and pressures* (Pang et al., 2020).

Artificial photosynthesis is currently unable to compete with fossil fuels in terms of cost, although some recent research shows that the efficiency is higher than with natural photosynthesis:

> *Using the tools of synthetic biology, a bioengineered bacterium, Ralstonia eutropha, converts carbon dioxide from air, along with the hydrogen produced from such catalysts of the Artificial Leaf, into biomass and liquid fuels, thus closing an entire artificial photosynthetic cycle. The Bionic Leaf operates at solar-to-biomass and solar-to-liquid fuels*

efficiencies that greatly exceed the highest solar-to-biomass efficiencies of natural photosynthesis (Dogutan & Nocera, 2019, p. 3143).

Researchers are thus looking for different ways to maximize efficiency:

An 'artificial tree' is a device that mimics the processes used by biological plant life to absorb CO_2 from the atmosphere. In nature, plants combine CO_2 from the atmosphere with water from their sap chemically, forming various hydro and oxy-hydrocarbons. However, in the case of artificial trees, the output from the 'tree' is a stream of essentially pure CO_2 at high pressure, ready for sequestration. Klaus Lackner, a physicist at Columbia University, is working on a filter that can do just that. Lackner envisions artificial trees small enough to fit in a shipping container but large enough to capture a ton of CO a day (Quader & Ahmed, 2017, p. 91).

It should be pointed out, however, that the cost-saving argument should not be the prevailing rationalization in the development of some ecologically sound solutions.

7.6.2 Various Contemporary Visions of Energy Use (Reuse)

To reduce our carbon footprint by electrifying society, massive deployment of renewable energy sources has become mandatory. This requires grid-tied energy storage system to balance energy production and consumption demands, to help grid stability.

There are many opportunities and challenges related to the technical and commercial development of such energy storage systems as grid components. These include batteries and superconductors, cryogenic storage, battery system design, inverters, safety considerations, etc. Technological advancements in this field include the following:

- Solar micro-inverters – using wide bandgap semiconductors to optimize their design
- A second life for EV batteries – maximizing their utilization by assembling them into reconfigurable battery systems
- Using carbon-ion supercapacitors – to achieve large-scale, high-longevity, and high-power density energy storage
- Building safety and reliability into battery energy storage systems – from initial design and creation, through manufacturing and operational life, to end-of-life recycling

7.6.3 Carbon Capture and Storage

"Bioenergy with carbon capture and storage (BECCS) is a carbon-negative technology that combines sustainable bioenergy conversion with CO_2 capture and storage (CCS). It is one of the more prospective large-scale carbon removal technologies. A challenge in artificial photosynthesis is to use cheap and environmentally friendly compounds. Many components currently proposed for use in artificial

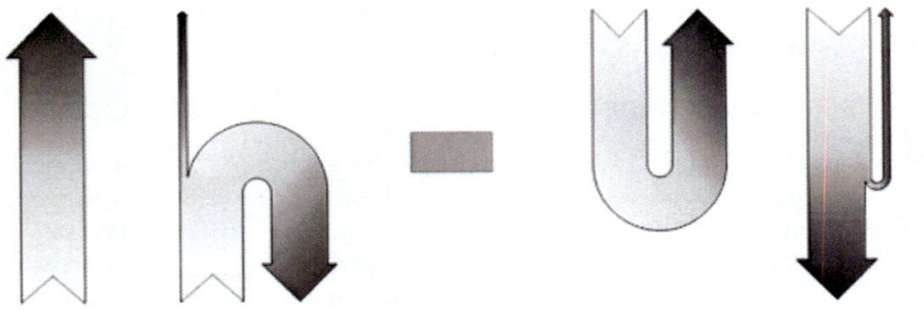

Fig. 6.20 Bioenergy with carbon capture and storage (BECCS)

photosynthetic systems are expensive, toxic, inefficient, or nondurable" (Quader & Ahmed, 2017, p. 91) (Fig. 6.20).

BECCS, the process of extracting bioenergy from biomass and capturing and storing carbon, thus removing carbon from the atmosphere (Obersteiner, 2001), is thus one of the steps in combating climate change, despite the many issues still associated with it.

Other authors also agree on the environmental impact:

> *Bioenergy with Carbon Capture and Storage (BECCS) features heavily in the energy scenarios designed to meet the Paris Agreement targets, but the models used to generate these scenarios do not address environmental and social implications of BECCS at the regional scale* (Donnison et al., 2020, p. 586).

7.6.4 Energy Reuse

Energy recycling is the *energy recovery*[31] process of utilizing energy that would normally be wasted, usually by converting it into electricity or thermal energy. Undertaken at manufacturing facilities, power plants, and large institutions such as hospitals and universities, it significantly increases efficiency, thereby reducing energy costs and greenhouse gas pollution simultaneously. The process is noted for its potential to mitigate global warming profitably. This work is usually done in the form of combined heat and power (also called cogeneration) or waste heat recovery. Cogeneration or combined heat and power (CHP) is the use of a heat

[31] *Energy recovery* includes any technique or method of minimizing the input of energy to an overall system by the exchange of energy from one subsystem of the overall system with another. The energy can be in any form in either subsystem, but most energy recovery systems exchange thermal energy in either sensible or latent form.

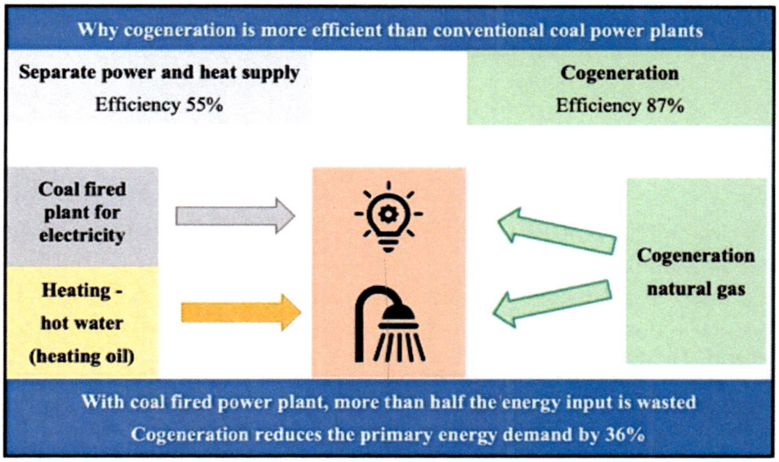

Fig. 6.21 Energy reuse

engine[32] or power station[33] to generate electricity and useful heat at the same time (Fig. 6.21).

Cogeneration is a more efficient use of fuel or heat, because otherwise wasted heat from electricity generation is put to some productive use. CHP plants recover otherwise wasted thermal energy for heating. This is also called combined heat and power district heating. Small CHP plants are an example of decentralized energy. By-product heat at moderate temperatures (100–180 °C, 212–356 °F) can also be used in absorption refrigerators[34] for cooling.

The supply of high-temperature heat first drives a gas- or steam turbine-powered generator. The resulting low-temperature waste heat is then used for water or space heating. At smaller scales (typically below 1 MW), a gas engine or diesel engine may be used.

Cogeneration was practiced in some of the earliest installations of electrical generation. Before central stations distributed power, industries generating their own power used exhaust steam for process heating. Large office and apartment buildings, hotels, and stores commonly generated their own power and used waste

---

[32] *Heat engine* is a system that converts *heat* to *mechanical energy*, which can then be used to do mechanical work.

[33] *Generating station* or *generating plant* is an industrial facility for the generation of electric power. Power stations are generally connected to an electrical grid.

[34] *Absorption refrigerator* is a refrigerator that uses a heat source (e.g., solar energy, a fossil-fueled flame, waste heat from factories, or district heating systems) to provide the energy needed to drive the cooling process.

steam for building heat. Due to the high cost of early purchased power, these CHP operations continued for many years after utility electricity became available. The cogeneration industry says that it can help mitigate climate change.

Waste Heat Recovery

Waste heat found in the exhaust gas of various processes or even from the exhaust stream of a conditioning unit can be used to preheat the incoming gas. This is one of the basic methods for recovery of waste heat. Many steel making plants use this process as an economic method to increase the production of the plant with lower fuel demand. There are many different commercial recovery units for the transferring of energy from hot medium space to lower one:

- *Recuperators.* This name is given to different types of heat exchanger that the exhaust gases are passed through, consisting of metal tubes that carry the inlet gas and thus preheating the gas before entering the process. The *heat wheel* is an example which operates on the same principle as a *solar air conditioning* unit.
- *Regenerators.* This is an industrial unit that reuses the same stream after processing. In this type of heat recovery, the heat is regenerated and reused in the process.
- *Heat pipe exchanger.* Heat pipes are one of the best thermal conductors. They have the ability to transfer heat hundred times more than copper. Heat pipes are mainly known in renewable energy technology as being used in *evacuated tube collectors*. The heat pipe is mainly used in space, process, or air heating; waste heat from a process is being transferred to the surrounding due to its transfer mechanism.
- *Thermal wheel* or rotary heat exchanger consists of a circular honeycomb matrix of heat absorbing material, which is slowly rotated within the supply and exhaust air streams of an air handling system.
- *Economizer.* In case of process boilers, waste heat in the exhaust gas is passed along a recuperator that carries the inlet fluid for the boiler and thus decreases thermal energy intake of the inlet fluid.
- *Heat pumps*: Using an organic fluid that boils at a low temperature means that energy could be regenerated from waste fluids.
- *Runaround coil* comprises two or more multi-row finned tube coils connected to each other by a pumped pipework circuit.
- *Particulate filters* (DPF) to capture emission by maintaining higher temperatures adjacent to the converter and tail pipes to reduce the amount of emissions from the exhaust.

8 Conclusion

If we want to address and solve the problem of energy production and consumption and, at the same time, to reduce greenhouse gas resources and act green, there are only two diametrically opposed options:

- Saving energy
- Using green energy sources

In between the two options, there is a range of intermediate, hybrid solutions, including more or less intense energy saving and the use of more or less environmentally friendly energy sources. Of course, modern technologies also play an important role in this respect, as they enable the reduction of greenhouse gas emissions through higher efficiency and cleaning of toxic gas emissions, which helps to partially neutralize their harmful impact on the environment. As mentioned, this can be achieved through a combination of general regulations at a global level and raising the awareness of individuals.

We are not saying that nuclear energy is green energy, but it is the next best thing in the given situation. If we want to have a reliable and affordable source in the medium-term future (by the year 2050–2060), the electricity mix will have to contain nuclear energy as well. While the idea of renewable and green energy sources may seem an attractive one, from the perspective of today's technology, it cannot actually provide solutions for the future or, much less, guarantee a sustainable way of producing energy.

Chapter 7
A View to the Future and Human Awareness

Abstract This chapter focuses on already-existing, alternative, under-researched, or unaffordable solutions, such as nuclear fusion (the most promising future energy source), artificial photosynthesis, or various other modern visions of energy extraction. We also touch on topics such as the storage of energy as well as greenhouse and other gases, recycling and reuse of energy waste and its processing, storage or reuse of wastes from different sources or technological processes, and their impact on the environment. Theoretical ideas and knowledge from previous chapters are summarized, and we no longer deal so much with the concrete ecological problems themselves as with trying to suggest potential solutions based on critical thinking and rational judgment. In doing so, we focus in more detail on possible solutions for two of the most pressing problems in terms of sustainability and the environment:

1. The problem of using alternative, non-traditional forms of energy production and resources. (We provide an outline of the critical relationship toward the use of nuclear energy among its opponents and advocates and attempt to demonstrate how we can develop critical thinking and a critical attitude toward certain issues at a systemic and long-term level.)
2. The problem of transportation, the related stress on the environment, and the impact of transport on the ecological footprint. (Modern trends for a transition to using electricity in transport are highlighted. Electricity is still mostly obtained from fossil fuels.)

In both cases, we focus mainly on the causal links between energy production and energy consumption and the impacts of these processes on the environment and the ecological footprint. These issues are observed from a long-term perspective, from the point of view of sustainability, environmental awareness, and a possible change in the attitude of the individual and the society as a whole.

In the previous chapters, we fully grasped the scope of environmental issues associated with the protection of the planet's natural environment. We also learned about awareness-raising methods. In this chapter, we will attempt to link these two aspects, as environmental protection is both a technological and a sociological as well as anthropological issue.

B. Aberšek, A. Flogie, *Human Awareness, Energy and Environmental Attitudes*,
https://doi.org/10.1007/978-3-031-05871-4_7

We will no longer deal so much with the concrete problems themselves as with trying to suggest potential solutions based on critical thinking and rational judgment. In doing so, we will focus in more detail on possible solutions for two of the most pressing problems in terms of sustainability and the environment:

The problem of using "alternative," non-traditional forms of energy production and resources. We will highlight the critical relationship toward the use of nuclear energy among its opponents and advocates and attempt to demonstrate how we can develop critical thinking and a critical attitude toward certain issues at a systemic and long-term level, with emphasis on environmental issues and the ecological footprint. One of the facts that need to be highlighted at the very beginning is that radioactive elements are elements, which are naturally present in our environment. They are not man-made. *Natural radioactivity*[1] is not a product of human activity and cannot be avoided. From a purely environmental point of view, humans simply refine and concentrate these naturally present elements (as we do with many other materials, for example, extracting precious metal ores from copper mines, etc.). We could say that this is a transition from global to local: global low-level radioactive "pollution" is converted into local high-level radioactive "pollution." And yet, nothing has changed globally; the amount of total radioactivity remains constant. The question is, how to exit the current environmental crisis, and what could be a medium-term solution (until 2050/55)? There is probably only one answer – nuclear energy.

The problem of transport, the related stress on the environment, and the impact of transport on the ecological footprint. We will highlight the modern trends for a transition to the use of electricity in transport. The process is reverse to the one described above: from a global problem of "even" pollution (e.g., exhaust gases of internal combustion engines), we are trying to concentrate locally all emissions to electricity production. We should not forget that electricity is still mostly obtained from fossil fuels, if we disregard hydropower plants and nuclear reactors (at least until the year 2050/55).

In both cases, we will focus mainly on the causal links between energy production and energy consumption and the impacts of these processes on the environment and the ecological footprint. These issues will be observed from a long-term perspective, from the point of view of sustainability, environmental awareness, and a possible change in the attitude of the individual and the society as a whole.

This chapter, therefore, contains our contribution to the discussion about future energy and what paths we should take to preserve our way of life and at the same time to protect the environment. In the first part, the focus is on possible future scenarios for the harnessing of nuclear power and, in the second part, on the problem of transport and its impact on our environment.

---

[1]The phenomenon of spontaneous and continuous as well as uncontrollable disintegration of an unstable nucleus accompanied by the emission of active radiations is called *natural radioactivity*. The substance that exhibits radioactivity is called a radioactive substance. e.g., uranium, thorium, radium, etc. Man did not invent the process of radioactivity; it was there, existing in the universe since time immemorial.

1 Problem 1: The Green Deal and Nuclear Power

Many potential solutions for protecting our natural environment, related to energy and energy sources, were discussed in previous sections. They could be divided into short-term (approximately until the year 2030), medium-term (until 2050), and long-term solutions, which extend into the more distant future. If we want to have constant and sustainable sources of green energy supply in the medium term, the only logical green solution could be nuclear energy, despite all the concerns and oppositions (Wealer et al., 2018; Wenisch et al., 2007). We will not focus on the economic aspects of nuclear power, because, as mentioned, nature has long since paid for these. Our focus here will be only on environmentally acceptable solutions. To begin with, let us focus briefly on the nuclear energy industry. Important concepts in this field include updating old fission reactors (short-term solution), Generation III (IV) systems (medium-term solution), and, finally, Generation IV fission reactors and nuclear fusion (long-term solution). While there is enough theoretical information available on both fission and fusion energy from research institutions, we attempt here to provide a broader perspective and examine how much is fiction and what these concepts could mean in some *future reality*, which is upon us to decide on *now. We will seek to polemicize and critically assess certain short- and medium-term issues, claims, and concerns and to comment on the issues logically and critically (LCC).* The arguments used for the interpretation and critical judgment fall under two categories:

- Logical, scientific arguments
- Arguments of the type "pathos" and "ethos", which are used primarily by pseudo- and fringe scientists

The basic starting point in presenting this problem (Problem 1, nuclear power) and the next one (Problem 2, transportation) is not in the construction and justification of scientific considerations, but rather in demonstrating how we need to build the awareness on the basis of critical thinking and concrete examples as well as concrete theses of pseudo-scientists and even those scientists who only tell us a partially embellished truth with the intent of manipulating arguments to deceive the public.

1.1 *Critics: Do We Need Nuclear Power to Meet the Challenge of Climate Change?*

The principal challenges today are sustainable development and climate change and how society can decide on a fair distribution of resources and energy as well as a radical reduction of ecological footprint. It is important to understand that the decay of CO_2 in the atmosphere is a slow process. Therefore, the impact of today's CO_2 emission reduction will be visible only in the coming decades. If we want to prevent severe climate change, there is no time to waste: *CO_2 reduction needs to be cheap,* *effective and implemented without delay. Can nuclear power be part of such measures?*

LCC[2]: Yes. In order to maintain today's way of life in the future, nuclear power is the only short- and medium-term solution currently available. Discussions about the economic effects of nuclear power are completely irrelevant and unjustified. Nature has long since paid the price.

1.2 Nuclear Fission

Experience with nuclear power and nuclear power plants (NPPs) does not suggest it could fulfill all ecological and sustainable requirements: enormous sums of money have been invested in the development and continuous improvement of nuclear energy technology since the 1950s. However, with a share of hardly 7%, nuclear energy is not substantially contributing to world energy demand.

LCC: This is a pathos-type, pseudo-scientific argumentation, ripped out of context: we learn nothing about what this means in terms of the whole (global) energy production. The share of nuclear energy contributing to world energy demand can grow if needed.

Reasons for this type of argumentation have been continuous technical and safety-related problems:

- The first generation of mostly small prototype nuclear reactors was characterized by many teething problems.
- Only the second generation – these are most of today's operating NPPs – could gain a short construction boom in the 1980s. Nuclear was cut off basically for good by the accident in Three Mile Island, USA, in 1979, which resulted in a total construction stop for NPPs in the United States. The severe reactor accident in Chernobyl, Ukraine, in 1986 and the Fukushima nuclear disaster in March 2011 caused a large part of the world population to become more sensitive to the dangers of nuclear power. People's trust in nuclear power and the awareness of world citizens about the existence of "peaceful" nuclear technology is low. This fact is confirmed every year by polls, which show that the world perceives nuclear power to be more of a risk than an advantage, moreover, most world citizens would either reduce or keep the current share of nuclear energy.

For example, Foratom, the European trade agency for nuclear industry, is heavily promoting a nuclear renaissance as a powerful tool to minimize CO_2 emissions. Foratom might be facing a tough job: at present, 439 NPPs worldwide generate 372-GW electrical energy, and 31 NPPs are under construction, most of which are in Asia and only one in Europe. During 2006, two new NPPs were connected to the grid, and six were permanently shut down (IAEA PRIS, 2007). Forecasts by IAEA, IEA, and OECD indicate no dramatic change, but only a slow growth of nuclear power by

[2]LCC – Logical (Scientific) Critical Comments.

2030 – with a net growth rate of only 600 MW annually. *With such a slow growth rate, nuclear power cannot play an important role in future world energy production.*

The nuclear industry in recent years, after the Fukushima accident, is improving its image and presence in the public discussion and increasing its marketing activities, but its actual capacity for the construction of new power plants is very limited. Due to decreased economic activity, financial losses, and a process of consolidation, only a handful of companies are still on the market. Currently, this industry must tackle the lack of skilled personnel and production capacities for its very specialized products. A substantial comeback in the shape of newly constructed plants before 2030 can be excluded.

1.3 Green Peace Perspectives – Nuclear Power and CO_2

In proving or disproving this argument, we will rely on a traditional way of reasoning, based on Aristotle's three modes of persuasion: logos (logical argument), pathos (appealing to the emotions and the sympathetic imagination, as well as to beliefs and values), and ethos (an appeal to the authority or credibility of the presenter/speaker). We will present Logical Critical Comments (LCC) for each individual claim, distinguishing them as arguments of the type logos, pathos, or ethos. In general, it can be said that logos arguments are predominantly scientific arguments, while pathos- and ethos-type arguments are mostly pseudo-scientific arguments. We characterize all arguments as either "pro" or "contra." In our argumentation, we will only pursue the benefits of the natural environment, not focusing on economic and other aspects. Our starting point is the fact that the situation today has been created by humans through our adaptation of the natural environment to meet our narrow interests (to say that we have done so to "meet our *needs*" would be unfair), which has caused a lot of damage to the Earth's ecosystem. Only human beings are responsible for this. Now, we must pay for what we did to nature, and give back what we took from it. Therefore, from an environmental point of view, the economic argument is clearly a pathos-type argument, used to justify certain actions and create an excuse for not paying the due, *which is unacceptable.*

Contra

Let us begin with arguments from those who are opposed to using nuclear power:

- Nuclear energy is not CO_2-free if the whole uranium fuel cycle is taken into consideration. Using current uranium ore grades (~2% concentration) results in 32 g of CO_2 equivalent (CO_2eq) per kWh of nuclear electricity (kWhel) in Germany. In France, it is only 8g/kWhel, while it is higher in Russia and in the USA, 65 g and 62 g respectively. One reason for this is the quality of uranium: the lower the grade, the more CO_2. A substantial increase in nuclear electricity generation would require the exploitation also of lower-grade uranium ores and thus would increase the CO_2 emissions up to 120 g CO_2eq/kWhel, which is much more than other energy technologies: natural gas co-generation, 50–140 g CO_2eq/kWhel; wind power, 24 g; hydropower, 40 g; and energy conservation, 5 g CO_2eq/kWhel) (OEKO, 2007).

LCC: Logos- and pathos-type argument. The argument is taken out of context, because no reference is made to what this means in the light of the world's total energy production. We should not forget that CO_2 is ultimately a useful gas, essential for the life of plants, which use it as food to create oxygen, which is essential for the existence of life on earth.

- Not to be forgotten is that uranium mining destroys the landscape and contaminates the soil, air, and water in the mining regions, and each ton of uranium creates several tons of tailings.

LCC: A pathos-type argument. Such pseudo-scientific argumentation also presents matters out of context; we learn nothing about what this means in terms of the whole (global) energy production. Everything people do destroys the environment and pollutes the earth, air, and water. If we didn't want this to happen, we would have to change our way of life, and that is something we are not ready for. Human activity causes environmental degradation through deforestation, urbanization, construction of cities and roads, not to mention coal mining or oil extraction (e.g., fracking).

- A very dangerous legacy is the so-called back end of the nuclear fuel cycle, the nuclear waste. Spent nuclear fuel is dangerous for the environment and for humans, and cannot be neglected for millions of years. Potential sites chosen for the repositories for highly active waste are usually rejected by the local population – provided public debate is not oppressed by the state.

LCC: This is a pathos-type, pseudo-scientific argumentation, ripped out of context: we learn nothing about what this means in terms of the whole (global) energy production. An analogy could be drawn to any kind of man-made waste – but people do not perceive this as a problem. Radioactive waste is a minor part of all waste, compared to municipal waste or waste produced by coal-fired power plants!

- The plans for Generation IV reactor systems intend to achieve the reuse of spent fuel. However, final repositories will still be needed.

LCC: This is a pathos-type, pseudo-scientific argumentation. No facts support this argument; it is pure speculation.

Let us now consider in more detail the problem of next-generation nuclear reactors.

1.4 Generation IV Nuclear Reactors

Research institutions and R&D departments of nuclear companies have developed new, Generation IV nuclear reactors, which, however, will not be in use before the year 2035 (Wenisch & Prauhart, 2006). Generation IV[3] breeder systems are a new

[3]Under the title Generation IV reactor concepts, the nuclear industry promises new systems as a response to the upcoming scarcity of fissile material (uranium), the issue of proliferation, and the requirement of safe repositories for high-level nuclear waste.

edition of the Fast Breeder concept, which would mean a continuation of the plutonium economy on a scale unseen until now. Another track being investigated is the thorium fuel route. This would be the continuation of the High Temperature Reactor development: a prototype was operated in the 1970s for the first time in Germany but never reached stable working conditions before the reactor was finally shut down 20 years later. Only one very small reactor of this type is in operation today – in China. The nuclear industry promotes its potential to be part of the solution of all energy problems of our planet. The nuclear industry promotes its potential to be part of the solution of all energy problems of our planet, in the form of new reactors, new fuel cycle systems, and, last but not least, by providing the world with a practically infinite energy source – nuclear fusion. During more than 50 years of research, construction, and operation of nuclear facilities, the nuclear industry has created a lot of problems without a solution (safety of nuclear installations, nuclear waste, radioactive emissions, etc.). However, they are still claiming that future nuclear energy will be better than the previous systems.

LCC: This is a pathos-type, pseudo-scientific argument: speculations and predictions based on age-old experiments. This is not logical argumentation, only conjecture and pseudo-prediction!

1.5 Reality

Generation IV reactor systems consist of reactors for energy production plus reprocessing facilities. Part of the systems are also plutonium breeders, which are designed to produce electricity and at the same time "breed" new plutonium. The reprocessing facilities separate fissile material from spent fuel to manufacture MOX fuel. This "recycling" of course generates new nuclear waste.

LCC: Another pathos-type, pseudo-scientific argument: all processes produce certain by-products that are not environmentally friendly, including renewable energy sources such as wind or solar; this is not logical argumentation, only conjecture and pseudo-prediction!

Vast quantities of highly toxic materials like plutonium and uranium isotopes would be transported around the world like oil or coal, posing a considerable accident risk as well as becoming a target for sabotage and terror.

LCC: A pathos-type, pseudo-scientific argument. What is under discussion here is not the transportation of materials itself, but especially the quantities; there is a significant difference between transporting a trainload of radioactive materials, or just a van.

Moreover, Generation IV reactor systems will probably enhance the possibilities to stash away fissile material, because the amount needed for a nuclear explosion is small, between 10 kg and 50 kg of uranium (^{235}U) or 5 kg of plutonium (^{239}Pu), depending on the construction. The efforts of the nuclear industry to construct more and more nuclear facilities and provide fissile material for civil purposes will probably outweigh the non-proliferation efforts of the IAEA. New reactors are expected to have simpler and cheaper designs. However, due to the extreme conditions under which those reactors are operating (extremely high temperatures, pressures, etc.), Generation IV reactors need even more sophisticated safety systems.

LCC: A pathos-type, pseudo-scientific argument. This is only speculation. Referring to safety systems is a powerful pathos-type argumentation.

It is widely assumed that Generation IV systems will not be commercially available before 2030, and there is no indication that these reactors will make nuclear power leave the league of the most expensive sources for electricity generation.

LCC: Again, this is a pathos-type, pseudo-scientific kind of reasoning, a negative campaign based on false arguments. (The costs of nuclear fusion reactors are comparable to the costs of thermal power plants and fission nuclear reactors. These range between 1.5 €/W and 3 €/W. However, they do not even take into account the costs of environmental pollution and do not include fuel costs, operating costs, and decommissioning costs, which add up to around 3 €/W, so the cost of today's power plants is somewhere between 4.5 €/W and 6 €/W.)

1.6 Background

We are currently witnessing the discussion whether and under which conditions to allow the large commercial reactors from the 1970s, 1980s, and 1990s (Generation II) to operate longer than the customary lifetime span of 30 years. At the same time, new reactors (Generation III) are being introduced – evolutionary designs developed from Generation II, without drastic changes. Generation III NPPs are already in operation in Japan and in construction in the European Union (EU). According to many scientists, the utilization of these reactors is limited in time because in their opinion, uranium reserves (uranium being the source material for nuclear fuel for most Generation III reactors) will dry up within the next three decades.

LCC: The lifespan of these reactors is 30–45 years. Irrelevant argument!

Due to these dismal prospects, the Generation IV International Forum (GIF) was founded in 2000. Until recently, it was composed of 10 countries (Argentina, Brazil, Canada, France, Japan, South Africa, South Korea, Switzerland, the United Kingdom, and the United States) as well as the EU via Euratom. At the end of 2006, China and Russia also joined the initiative. Membership in this international forum commits participating countries to support long-term research efforts. This also includes countries that are opposed to nuclear power. In 2001, the IAEA initiated International Projects on Innovative Nuclear Reactors and Fuel Cycles (INPRO). It is funded through the IAEA budget. GIF and INPRO have agreed to formalize cooperation at the technical level (Hirsch et al., 2005). As of February 2007, 28 countries or entities are members of INPRO.

The message of GIF and INPRO to media, politicians, and the population is as follows: Generation IV means a safe, economically competitive, and proliferation-resistant power source without the problem of increasing greenhouse gas emissions.

Generation IV is even presented as sustainable, a label which is usually – and with good reason – reserved for renewable energy sources and conservation. The fact that none of the six reactor concepts selected for development fulfils all Generation IV aims is usually not mentioned (Hirsch et al., 2005).

Generation IV does not only stand for the development of new reactor types. It is a comprehensive framework program for an international research cooperation effort to develop new nuclear energy systems – from resource extraction to final nuclear waste disposal. GIF has laid down its research needs and formulated eight development goals, which should fulfill four political goals: *sustainability, economic viability, safety and protection against acts of terrorism* (Schulenberg et al., 2004). The main general goals are as follows:

1. Sustainable energy production with long-term available systems based on an effective usage of fissile material with the lowest possible contamination of the air.
2. Improvement of the protection of humans and the environment by minimizing nuclear waste and the decay heat of the waste, as well as reducing long-term radiotoxicity.
3. Clear economic competitiveness with other energy sources.
4. Financial risk comparable to the financial risk of other energy projects.
5. Improved safety and reliability.
6. Lower probability of occurrence and minor consequences of nuclear core damage.
7. No need for emergency measures outside the nuclear installation.
8. The theft of weapon-grade fissile material should be made more difficult or impossible, plus best protection against terror attacks.

LCC: As can be deduced from the above, only two of the goals are related to environmental concerns (sustainability), two are related to economic reasons, and the last three, to safety, or actually, to human stupidity. Technology opens the doors, but what we do with it – this is a decision made at the level of man as a social being!

1.7 Generation IV and the Nuclear Industry_Contra

By all appearances, the main goal is to save the sinking ship by trying to win back the trust in nuclear power among the population. Obviously, facelift uses climate change as one of the most important arguments to demonstrate the need for nuclear power. The nuclear industry wants to give the new generation of reactors the image of being sustainable, economically viable, safe, reliable, and terror-resistant. Many respected institutions

LCC: (e.g., the Massachusetts Institute of Technology MIT – LCC – ethos-type argument)

consider GIF's goals as unrealistic. The following chapter examines how realistic these ambitious GIF Technology Roadmap (2002) goals are.

LCC: Speculation made to facilitate the path for pathos-type arguments. The solutions are pseudo-scientifically manipulated in order to create a pathos atmosphere.

1.8 Eight Claims and Eight Nuclear Daydreams

Sustainability is a concept that not only considers a comprehensive human-ecological context but also a broader time horizon. According to the Brundtland Commission 1987, "sustainable development is development that meets the needs of the present without compromising the ability of future generations to meet their own needs." Sustainability could be defined as making use of the natural system in such a way that its primary features are maintained in the long term and can be handed over to future generations as unchanged as possible.

2 First Goal – Generation IV and Sustainability

Generation IV nuclear energy systems will provide sustainable energy generation that meets clean air objectives and promotes long-term availability of systems and effective fuel utilization (NERAC, 2002)[4] is the first goal GIF is pursuing. But the uranium extraction itself generates 80% of today's radioactive waste (by mass, not by radioactivity).

LCC: Still, this radioactivity is natural – it does exist in nature, and man is simply concentrating it, reducing the mass of the energy source and, in turn, the quantity of waste, which is not the case with fossil fuels, where we do not concentrate the energy value (or indeed, coke production (from coal) is an example of this kind of "artificial concentration" in terms of fossil fuels).

To produce 1 ton of nuclear fuel, which is used in light-water reactors, several thousand or tens of thousands of tons – depending on the uranium content – of uranium ore must be extracted

LCC: The ore would be extracted from natural sources – and this is the data that needs to be compared to the quantity of energy that may be extracted from it.

The amount of radioactive tailings left behind in the uranium mine area is of corresponding volume. For example, the affected regions of New Mexico (USA) and Wismut (former GDR) must cope with more than 100 million tons of radioactive waste from uranium extraction on the surface (Hirsch & Weish, 2007).

LCC: Let us compare this, for example, with coal and the amount of energy gained from coal mining.

[4] https://www.hsdl.org/?view&did=883

Even if all other problems were solved, energy generation with nuclear energy systems (what about nuclear fusion?) could only be advantageous in the short term, and not in the longer term, because of the waste issue.

LCC: The matter under discussion is not the entire excavated quantity, but rather the part of it gained by enrichment, while the rest of the excavated material returns to nature.

Radioactive waste, which is created by the "peaceful" use of nuclear power, represents an extraordinarily high long-term threat potential.

LCC: The same can be said of any other type of waste, such as coal ash (coal combustion residuals).

It cannot be neglected for millions of years, which is unique in the industrialized society (Hirsch & Weish, 2007). Many geologists

LCC: (these are ethos-type arguments; who exactly are these "many geologists"?)

warn against the waste management option highly regarded mainly in Europe: the deep underground repository, which will be made inaccessible with barriers of concrete. The main problem is that it is simply impossible to conduct a safety case for the necessary long time periods of millions of years.

LCC: But what happens to radioactivity in millions of years; what about the half-life of elements?

At this point, science reaches its limit to make predictions (Hirsch & Weish, 2007).

LCC: This is a pathos-type argument that can only be made by a pseudo-scientist. Science is certainly capable of calculating, by how much the radioactivity will decrease in millions of years.

The disposal site would have to be protected continuously against a variety of threats including water ingress, overheating, sabotage, and theft of the waste for abuse, which is clearly impossible. Not only earthquakes, distortions, and volcanism but also future ice ages might pose a threat because glaciers can dig up and remove enormous amounts of rock.

LCC: Again, a pathos-type pseudo-scientific argument. Science knows how to predict, i.e., calculate what would happen in, for example, an ice age. Science knows where volcanoes are located, as well as lithospheric plates' fractures, which are related to global earthquake areas. All nuclear power plants have been built while bearing in mind these parameters – so why not radioactive waste disposal sites?

Gradually, however, a shift away from nuclear power is taking place. The Generation IV initiative attempts to reverse this shift by making nuclear energy attractive and presenting it as sustainable and CO_2-free, labels usually – and with justification – reserved for renewables. This strategy will help the nuclear industry and nuclear research institutions survive (Hirsch et al., 2005).

3 Second Goal – Generation IV and Nuclear Waste

The second goal, *Generation IV nuclear energy systems will minimize and manage their nuclear waste and notably reduce the long-term stewardship burden, thereby improving protection for the public health and the environment* (NERAC, 2002), is a high-strung promise and would of course require a large amount of research funds.

LCC: This is a pathos-type argument. Such predictions are made with no proof to support them.

The announced reduction of nuclear waste has to be properly examined. According to GIF, the *closed fuel cycle* is celebrated as a major advantage of Generation IV concepts. A system with a closed fuel cycle is regarded as more effective and sustainable (Hirsch et al., 2005). A 2003 study by the US Massachusetts Institute of Technology found that the fuel cost with a closed fuel cycle, including waste storage and disposal charges, is about 4.5 times the cost of a once-through fuel concept.

LCC: Again, a pathos-type argument: comparing the incomparable (i.e., efficiency and sustainability vs. costs).

Therefore, it is obviously not realistic to expect that there ever will be new reactor and fuel cycle technologies that simultaneously overcome the problems of cost, safe waste disposal, and proliferation

LCC: Pure conjecture without any argumentation, not even pseudo-scientific arguments.

The long-term waste management benefits of advanced closed fuel cycles, involving reprocessing of spent fuel, are indeed not outweighed by the short-term risks and costs, including proliferation risks [MIT 2003].

To put reprocessing in a positive light, the nuclear industry has started to call it recycling. Reprocessing is not new; it has been practiced for quite a long time. The wording might give the impression of a closed cycle which – in the name of sustainability – does not use up resources or generate nuclear waste. However, this is a misperception.

LCC: This kind of reasoning (pathos-type argumentation) is basically just wordplay. Recycling can be either complete, or partial, and recycled materials may be used for various purposes. This kind of reinterpretation of the meaning of terms (words) is only a means of distracting attention from what is essential.

The fuel chain (a more appropriate wording than "fuel cycle") always requires fresh uranium, which is not renewable and can be produced merely at the high price of enormous environmental damage. And that is not the only inconvenient truth: during operation of the reactor, atomic nuclei with a high mass number (e.g., ^{235}U) are continuously split in the nuclear fuel into nuclei with low mass numbers with a high neutron surplus. The conversion of fissile material into fission products releases energy. The fission products are often unstable and therefore radioactive. The gaseous fission products, like the radioactive isotopes of the noble gases argon, krypton, and xenon (mainly the radiologically important isotopes ^{133}Xe and ^{85}Kr), and the volatile radioactive iodine isotope (^{131}I) can partially escape – depending on the tightness of the fuel rod claddings – to the atmosphere and can be traced in the surrounding of nuclear installations. Solid fission products (e.g., radioactive isotopes of strontium and cesium) have to be stored as radioactive waste. However, even here, a small share escapes in the form of radioactive aerosols (^{134}Cs, ^{137}Cs, ^{90}Sr).

LCC: A pseudo-scientific claim, which is at the same time incorrect, misleading, and inconsistent!

Not all fissile materials in a fuel rod can be split during irradiation. To make use of the still-usable share of the original fissile material – instead of storing it unused in the final repository – it has to be separated by reprocessing. At first, this complex and hazardous process appears to reduce the radioactivity of the waste. However, by making use of this additional fissile material from reprocessing in fission reactors, still more fission products are generated that have to be stored as waste in final repositories. Moreover, the end of the nuclear age is going to be deferred, because more nuclear facilities have to be built and finally have to be decommissioned (leading to the production of "decommissioning waste"). The "recycling" of spent fuel rods results in a dangerous concentration of radioactive fission products. This results in a larger volume of total waste, partly with a higher activity concentration, in the long-term.

LCC: A pathos-type argument; predictions with no proof to support them; a non-scientific and even non-pseudo-scientific approach.

In addition to the already-discussed fission products, actinoids are also produced during reactor operation, among them plutonium, which can be used for nuclear weapons and nuclear fuel. The action ides can be degraded by neutron bombardment, a fact that the systems of Generation IV are destined to make use of for the reduction of the amount of waste already inside the reactor.

LCC: A pathos-type argument; predictions with no proof to support them; a non-scientific and even non-pseudo-scientific approach.

Today around 10,000 tons of spent nuclear fuel is generated per year (Hirsch & Weish, 2007). Each ton not only contains fission products that have to be stored in some final repository but also several kilograms of plutonium and other actinoides (actinides). In a reprocessing plant, these materials are separated. The plutonium is either mixed with non-enriched uranium to produce mixed oxide fuel (MOX fuel) and used again to fuel reactors (whereby the plutonium share is around 5%) or for building nuclear weapons. The separated uranium is enriched in the isotope ^{235}U in enrichment plants to reach enrichment up to 3–5%. This uranium can then also be re-used in nuclear reactors as LEU (low enriched uranium) reactor fuel, while enrichment to 20% and more provides HEU (highly enriched uranium) applicable for building nuclear weapons (for an advanced fission explosive device, uranium is usually enriched to 90% or more).

LCC: Another pathos-type argument: in some way or another, human beings have always created weapons. Technology is not at fault here; it is not the gun that kills, but the person holding the gun!

The depleted uranium, unusable for reactors, is mainly used for the construction of airplanes and penetrating ammunition. The unusable radioactive fission products are "for the time being" put in interim storage since the question of final disposal has not yet been solved. What would be needed is practically the complete separation of all long-lived nuclides, so that the remaining waste needs to be stored safely for only a shorter time. A separation efficiency of 99% as achieved so far is not really sufficient when taking into account the amount of waste that is actually generated! Reprocessing also leads to high emissions in gaseous and liquid forms, which still contain radioactive substances in spite of off-gas and wastewater treatment procedures.

It would be necessary to develop some "super-reprocessing" technology (separation efficiency of 99,99%), which would, contrary to current methods, not damage the environment and not pose a catastrophic danger (Hirsch & Weish, 2007).

LCC: Misleading idea: talking about a separation efficiency of 99,99% would be illusionary. Currently, the highest levels of efficiency (reliability) exist in the aviation and nuclear industry, amounting in theory to between 99,95 and a maximum of 99, 98%.

As a measure to improve their image, the nuclear industry is now announcing the complete degradation of long-term toxic actinoides during reactor operation. For this purpose, fast neutrons are used in the reactor core. The sodium-cooled fast reactor should be able to reach the required 99,99% conditioning and reprocessing of the actinoides with the help of an adjusted reactor geometry and a closed fuel cycle approach (Schulenberg et al., 2004). What is omitted here: the unrecyclable fission products are still left over. Generation IV reactors are far away from the goal to successfully minimize and manage their nuclear waste (Hirsch et al., 2005). This is where the story ends: the operation of nuclear installations is probably never possible without the creation of radioactive waste. It is our opinion that nuclear technology cannot contribute to environmental protection.

LCC: It is a fact that no kind of production is possible without creating a certain amount of waste – and this goes for the production of energy as well. The only thing needed is to perform an analysis on which kind of energy production generates the least waste and, especially, which kind is less harmful to the environment in terms of sustainability.

4 Problem 2: Transportation and Energy

Transportation activities are significant energy consumers, providing mobility to passengers and freight, which accounts for about 25% of world energy use and about 61.5% of all the oil used each year.

Transportation and energy are a standard physics application, where giving momentum to a mass (passengers, vehicles, cargo, etc.) requires a proportional amount of energy. The relationship between transport and energy is a direct one but subject to different interpretations because it concerns different transport modes, each having their utility and level of performance. There is often a compromise between *speed* and *energy consumption* related to the desired economic returns.

Passengers and high-value goods can be transported by fast but energy-intensive modes since the time component of their mobility tend to have a high value, which conveys the willingness to use more energy. Economies of scale, mainly those achieved by maritime transportation, are linked to low levels of energy consumption per unit of mass being transported, but at a slower speed. This fits relatively well freight transport imperatives, particularly for bulk. Comparatively, air freight has high energy consumption levels linked to high-speed services (World Energy Council, 2007).

The transportation market has a wide spectrum of energy consumption that are particularly impacted by three issues:

- The *price level and volatility of energy sources* that are dependent on the processes used in their production. Stable energy sources are preferred as they enable long-term investments in transportation assets. Volatile energy prices are not contingent on investments in transport technology.
- *Technological and technical changes* in the level of energy performance of transport modes and terminals. An important goal is thus to improve this energy performance because it is linked with direct economic benefits for both operators (lower operating costs) and users (lower rates).
- *Environmental externalities* related to the use of specific modes and energy sources and the goal to reduce them.

A trend that emerged since the 1950s concerns the growing share of transportation in the world's total oil consumption; transportation accounts for approximately 25% of world energy demand and about 61.5% of all the oil used each year. The impacts of transport on energy consumption are diverse, including on activities that are necessary for the provision of transport infrastructures and facilities:

- *Vehicle manufacture, maintenance, and disposal.* The energy spent on manufacturing and recycling vehicles is a direct function of vehicle complexity, the material used, fleet size, and vehicle life cycle.
- *Vehicle operation.* Mainly involves energy used to provide momentum to vehicles, namely, as fuels, as well as for intermodal operations. The fuel markets for transportation activities are well developed.
- *Infrastructure construction and maintenance.* The building of roads, railways, bridges, tunnels, terminals, ports, and airports and the provision of lighting and signaling equipment require a substantial amount of energy. They have a direct relationship with vehicle operations because extensive networks are associated with large amounts of traffic.
- *Management of transport operations.* The expenses involved in planning, developing, and managing transport infrastructures and operations require time, capital, and skill that must be included in the total energy consumed by the transport sector. This is particularly the case for public transit.
- *Energy production and trade.* The processes of exploring, extracting, refining, and distributing fuels or generating and transmitting energy also require power

sources. The transformation of 100 units of primary energy in the form of crude oil produces only 85 units of energy in the form of gasoline. Any changes in transport energy demands influence the pattern and flows of the world's energy markets.

Energy consumption has strong *modal variations*:

- *Land transportation* accounts for the vast majority of energy consumption. Road transportation alone is consuming, on average, 85% of the total energy used by the transport sector in developed economies. This trend is not uniform within the land transportation sector itself, as road transportation is almost the sole mode responsible for additional energy demands over the last 25 years. Despite a falling market share, rail transport, on the basis of 1 kg of oil equivalent, remains four times more efficient for passenger and twice as efficient for freight movement as road transport. Rail transport accounts for 6% of the global transport energy demand.
- *Maritime transportation* accounts for 90% of cross-border world trade, as measured by volume. The nature of water transport and its economies of scale make it the most energy-efficient mode because it *uses only* 7% of all the energy consumed by transport activities, a figure way below its contribution to the mobility of goods. Nevertheless, fuel consumption is still an important factor in maritime shipping, which is related to ship design (hydrodynamics), utilization level, operational speed, idle time (waiting at ports), and even weather conditions. For terminal operations, figures vary, but a container terminal usually has 70% of its energy consumption provided by fossil fuels (e.g., yard equipment) and 30% by electricity (e.g., portainers).
- *Air transportation* plays an integral part in the globalization of transportation networks. The aviation industry accounts for 8% of the energy consumed by transportation. Air transport has high energy consumption levels, linked to high speeds. Fuel is the second most significant cost for the air transport industry accounting for 13–20% of the total expenses. Technological innovations, such as more efficient engines and better aerodynamics, have led to a continuous improvement of the energy efficiency of each new generation of aircraft.

Further distinctions in the energy consumption of transport can be made between the mobility of *passenger* and *freight*:

- *Passenger transportation* accounts for 50–60% of the energy consumption derived from transportation activities. The private car is the dominant mode but has a poor energetic performance, although this performance has seen substantial improvements since the 1970s, mainly due to the growing energy prices and regulations. Only 12–30% of the fuel used by a car provides momentum, depending on the type of vehicle. There is a close relationship between rising income, automobile ownership, and distance traveled by vehicles. The United States has one of the highest levels of car ownership in the world with one car for every two people. Another trend has been the increasing rise in ownership of minivans, sport utility vehicles, and light-duty trucks for personal use and the corresponding decline in fuel economy. However, fuel consumption is impacted by diminishing returns, implying that higher levels of fuel efficiency involve declining marginal gains in fuel consumption. Also, the growth of vehicle-miles traveled is correlated with changes in energy prices, underlining an elasticity.

- *Freight transportation* accounts for 40–50% of energy consumption derived from transportation activities. Energy consumption is dominated by road transportation, which can account for 80% of domestic consumption. Rail and maritime shipping, the two most energy-efficient modes, have more marginal consumption levels. Coastal and inland waterways also provide an energy-efficient method of transporting passengers and cargoes. Because of these energy advantages, short sea shipping is being considered as a transport alternative and part of national transport policies of countries having an extensive coastline. The rationale for favoring coastal and inland navigation is based on lower energy consumption rates and the overall lower externalities of water transportation.

4.1 *Transport Emission and Environment*

The figures show emissions from individual modes of transport. In order to provide for a more vivid illustration – because science, too, needs to be simplified to such an extent that it is still sufficiently accurate but also understandable to the everyman, not just to experts, as only in this way will we be able to fight against pseudo- and fringe science – let us distinguish between only two types of emissions for now: (1) CO_2 *emissions*, the most widely known and most frequently considered greenhouse gas, and (2) emissions by any *other pollutant*. In the following section, we will discuss individual emissions in more detail. Some of the most relevant questions that arise regarding the above are, for example, how do emissions vary? Can driving be better than flying or travelling by boat? (Fig. 7.1).

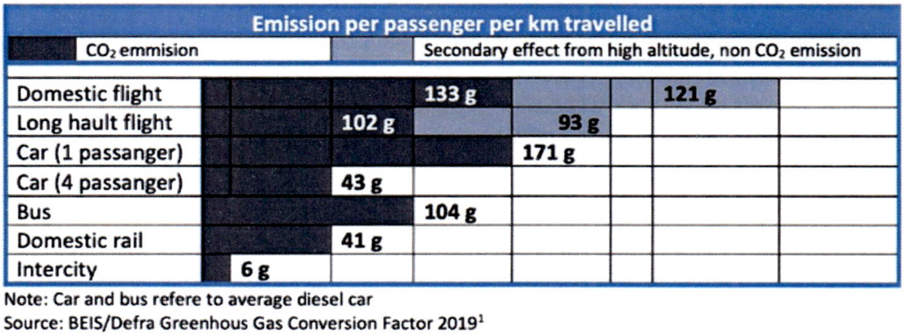

Fig. 7.1 Emission from different modes of transport

4.2 How Do Emissions Vary?

The warming effect of CO_2 and other non-CO_2 emissions, such as nitrogen oxides, differ according to where they are emitted. When they are released at high altitudes, they can make a significant difference to emissions calculations. The climate effect of CO_2 and non-CO_2 emissions from aviation is much greater than the equivalent from other modes of transport, as these non-CO_2 greenhouse gases formed at higher altitudes persist longer than at the surface and also have a stronger warming potential. But there is scientific uncertainty about how this effect should be represented in calculators. The EcoPassenger calculator[5] launched by the International Railways Union in cooperation with the European Environment Agency says it depends on the height the plane reaches. Train virtually always comes out better than plane, often by a lot. For example, a journey from London to Madrid would emit 43 kg (95 lb) of CO_2 per passenger by train, but 118 kg by plane (or 265 kg if the non-CO_2 emissions are included), according to EcoPassenger.

However, the margin between train and plane emissions varies, depending on several factors, including the type of train. For electric trains, the way the electricity they use is generated is used to calculate carbon emissions. Diesel trains' carbon emissions can be twice those of electric ones. Figures from the UK Rail Safety and Standards board show that some diesel locomotives emit more than 90 g of $C0_2$ per passenger per kilometer, compared with about 45 g for an electric intercity, for example. These data, however, are not unidimensional, because as mentioned earlier, the production and consumption of electrical energy are closely related, and we must always take into account the manner in which this energy was produced. For example, the source of the electricity can make a big difference if you compare a country such as France, where about 75% of electricity comes from nuclear power, with Poland, where about 80% of grid power is generated from coal.

4.3 Can Driving Be Better Than Flying?

Yes, if the car is electric – but diesel and petrol cars are also in many cases better options than flying, though it depends on various factors, particularly how many people they are carrying. According to EcoPassenger, a journey from London to Madrid can be done with lower emissions per passenger by plane, even accounting for the effect of high-altitude non-CO_2 emissions, if the car is carrying just one person and the plane is full. If you add just one more person into the vehicle, the car wins out. Coaches also score well. Travelling by coach emits 27 g of CO_2 per person per kilometer, compared with 41 g on UK rail (but only 6 g on Eurostar) – though again this will vary depending on how full they are and the engine type (Lowe, 1990).

---

[5] http://www.ecopassenger.org/

4.4 What About Travelling by Boat?

The BEIS has also put a figure on ferry transport – 18 g of CO_2 per passenger kilometer for a foot passenger, which is less than a coach, or 128 g for a driver and car, which is more like a long-haul flight. But ferries' ages and efficiency will vary around the world – and a ferry will not get you to America, although a cruise ship or ocean liner would. The cruise industry has long been under pressure to reduce environmental impacts ranging from waste disposal to air pollution, as well as high emissions, not only from travel but also from powering all the on-board facilities.

In conclusion, we wish to point out that the data, which may be obtained from different sources, are mostly one-dimensional, depending on who is the author of the viewpoint and on who paid for the research. Therefore, it is always necessary to check the sources and compare them to other similar ones. Today, in the age of the Internet and various search engines, this should not be a problem.

4.5 The Transport Energy Sources

4.5.1 Transportation and Oil Use

The extent to which conventional non-renewable fossil fuels will continue to be the primary resources for nearly all transportation fuels is subject to debate. But the gap between demand and supply, once considerable, is *narrowing*, an effect compounded by the peaking of global oil production. The steady surge in demand from developing economies, particularly China and India, requires additional outputs. This raises concerns about the capacity of major oil producers to meet this rising world demand. The producers are not running out of oil, but the existing reservoirs may not be capable of producing on a daily basis the increasing volumes of oil that the world requires. Reservoirs do not exist as underground lakes from which oil can easily be extracted. There are geological limits to the output of existing fields. This suggests that additional reserves need to be found to compensate for the declining production of existing fields. Reserve additions may not be enough to offset this growing demand, but technological improvements allowed to tap bitumen and oil shale reserves. However, extracting such reserves necessitates much energy and water. The production of a barrel of bitumen requires burning 10–20% of the energy content of the resulting crude oil in the form of natural gas.

Others argue that the history of the oil industry is marked by *cycles of shortages and surplus*. The rising price of oil will render oil recovery in difficult areas cost-effective. Deepwater drilling, extraction from tar sands, and oil shale should increase the supply of oil that can be recovered and extracted from the surface. But there is a limit to the capacity of technological innovation to find and extract more oil around the world, and the related risks can be very high. Adding oil extraction, distribution, and refinery capacity is a slow, complex, capital-intensive, and highly regulated endeavor. Carbon sequestration in the form of CO_2 capture and storage, if technically and economically viable, could enhance the recovery of oil from conventional wells and prolong the life of partially depleted oil fields well into the next century.

The main concern is the amount of oil that can be pumped to the surface daily, especially where major oil fields have reached peak capacity. Under such circumstances, oil prices are bound to rise in the medium to long term, sending significant price signals to the transport market. How the transport system will respond and adapt to higher energy prices is obviously subject to much debate and interpretations in terms of the scale and timing of the transition.

Almost all transportation modes depend on variations of the *internal combustion engine*, with the two most salient technologies being the *diesel engine* and the *gas turbine* because they are the linchpin of globalization. While ship and truck engines are adaptations of the diesel engine, jet engines are an adaptation of the gas turbine. Transportation is almost entirely reliant (95%) on petroleum products, with the exception of railways using electrical power. While the use of petroleum for other economic sectors, such as industrial and electricity generation, has remained relatively stable, the growth in oil demand is mainly attributed to the growth in transportation demand. What varies is the type and the quality of petroleum-derived fuel being used. While maritime transportation relies on low-quality bunker fuel, air transportation requires a specialized fuel with additives. Road transportation is highly fragmented, with 85% of the automobiles relying on gasoline, while 90% of the trucks rely on diesel.

It is worth having a closer look at the chemical combustion principle of hydrocarbons. For most internal combustion engines, gasoline serves as fuel, but other sources like methane (CH_4; gas turbines), diesel (mostly trucks), bunker fuel (for ships), and kerosene (turbofans of jet planes) are used. Gasoline produces around 46,000 Btu per kilogram combusted, which requires 16–24 kg of air. The energy released by combustion causes a rise in the temperature of the products of combustion. Several factors and conditions influence the level of combustion in an internal combustion engine to provide momentum and keep efficient operating conditions. The temperature attained depends on the rate of release and dissipation of the energy and the number of combustion products. Air is the most available source of oxygen, but because air also contains vast quantities of nitrogen, nitrogen becomes the principal constituent of the products of combustion. The rate of combustion may be increased by finely dividing the fuel to increase its surface area and hence its rate of reaction and by mixing it with the air to provide the necessary amount of oxygen to the fuel. The problem is that combustion in internal combustion engines is *imperfect* and *incomplete* for two reasons:

- First, the *fuel and the oxidizer are not pure*, causing imperfect combustion. Although the refining process provides a "clean" fuel, gasoline is known to have impurities such as sulfur (0.1–5%), sometimes lead (anti-knock agent being phased out), and other hydrocarbons (like benzene and butadiene). In comparison, air is composed of 78% nitrogen and 21% oxygen. Thus, other chemical components are part of the combustion process.

- Second, in part because of the first reason and in part because of the technology of the engine, *incomplete combustion emits other residuals.* Combustion in an engine occurs at an average rate of 25 times per second, leaving limited time for a complete combustion process. Besides carbon dioxide and water, a typical internal combustion engine will produce carbon monoxide (CO), hydrocarbons (benzene, formaldehyde, butadiene, and acetaldehyde), volatile organic compounds, sulfur dioxide (SO_2), particulates, and nitrogen oxides (NOx). These combustion products are the primary pollutants emitted into the environment by transportation.

In addition to the imperfect and incomplete combustion of hydrocarbons, three major factors influence the rate of combustion and thus emissions of pollutants:

- *Characteristics of the vehicle* (where technological improvements can play a role)
- *Driving characteristics* (where planning and regulation can play a role)
- *Atmospheric conditions* (The internal combustion engine, mostly due to friction, converts less than a third of the energy they consume into momentum. For electric motors, this figure is above 80%.)

The following potential consequences can be noted:

- *Road.* Depending on their level of productivity, many economies could show remarkable resilience. The next phase would see changes in commuting patterns (e.g., ridesharing, carpooling), attempts to use public transit, rapid adoption of vehicles with high gasoline efficiency, and a search for other transport alternatives.
- *Rail.* Rail is two times more energy-efficient than trucking. In many cases, there could be a push toward the electrification of strategic long-distance corridors and the development of more efficient cargo-handling facilities. Thus, growing energy prices are likely to affect long-distance rail transportation differently depending on the geographical setting and the conditions of the existing system (IEA, 2019).
- *Air.* This mode could be significantly impaired, both for passengers and freight. Air transportation is a highly competitive industry, and the profit margins are low. Fuels account for about 20% of the operating expenses of an air carrier, but because most of the other costs are fixed, any variations in energy prices are reflected directly on airfares. A long-term increase in energy prices, reflected in jet fuel, is likely to impact discretionary air travel (mainly tourism), but air freight, due to its high value, may be less impacted. Technological developments are helping to maintain the competitiveness of air transportation with more fuel-efficient planes. However, it needs to be noted that pollution of the upper (ozone) layers of the atmosphere remains an issue, which directly affects the entire ecosystem.
- *Maritime.* This mode is likely to be relatively unaffected as it is the most energy-efficient, but fuel is an important component of a ship's operating costs. The response of maritime shippers over higher energy prices tends to be lowering speed (slow steaming), which may have impacts on port call scheduling. In the

long run, higher energy prices may indirectly impact maritime transportation by lowering demand for long-distance cargo movements and incite port calls at ports having the most direct and efficient hinterland connections. In addition, this context may favor the development of short coastal and fluvial services where possible. Although maritime transport could be considered environmentally friendly in comparison to other modes of transportation, all these ships produce some pollution, as in the majority of cases, they burn low-grade bunker fuel that contains up to several thousand times the amount of sulfur compared with diesel fuel used in automobiles.

4.6 Modal Variations and the Environment

4.6.1 Road Transportation

Many research estimates that gasoline and plug-in hybrids are likely to play in road transport an important role in the 2035 time frame (15–40% and 7–15%, respectively). Strictly speaking, the gasoline hybrids are more of a fuel-efficiency improvement than a new technology, placing new demands on the electricity grid. However, several important parts of the pathway described below concerning batteries are also relevant to this technology, especially if it expands dramatically. On the basis of the AEF analysis, a significant market penetration of either fuel-cell or full-electric-battery vehicles is unlikely within the 2030 time frame.

The energy source with the lowest cost is usually preferred. The dominance of petroleum-derived fuels is a result of the relative simplicity with which they can be stored and used in internal combustion engines. Other fossil fuels (natural gas, propane, and methanol) can be used as transportation fuels as well but require a more complicated storage system. The main issue concerning the large-scale uses of alternative vehicle fuels is the significant capital investments required in distribution facilities as compared with conventional fuels. Another issue is that in terms of energy density, these alternative fuels have lower efficiency than gasoline and thus require a greater volume of onboard storage to cover the equivalent distance as gasoline propelled vehicle if performance is kept constant (Goldman & Gorham, 2006).

Alternative fuels in the form of non-crude oil resources are drawing considerable attention as a result of the non-renewable character of fossil fuels and the need to reduce emissions of harmful pollutants (Gilbert & Perl, 2008). The most prevalent alternatives being considered are as follows:

- *Biofuels* such as ethanol, methanol, and biodiesel can be produced from the fermentation of food crops (sugar cane, corn, cereals; often called first-generation biofuels) or biomass (such as wood and grasses; called second-generation biofuels). Their production, however, requires large harvesting areas that may compete with other types of land use. This limit is related to the capacity of plants to absorb solar energy and transform it through photosynthesis. This low productivity of the biomass does not meet the energy needs of the transportation sector. Besides, the production of ethanol is an energy-intensive process. Biodiesel can also be obtained from a variety of crops. The choice of biomass fuel will largely depend on the sustainability and energy efficiency of the production process. In general, biofuels are equal polluters as traditional, oil-based fuels.

- *Natural gas* is a more efficient and environmentally sustainable fuel for the transportation sector, namely, its compressed form. Although natural gas was used as a transportation fuel since the early twentieth century, its use remained marginal until the late twentieth century. It is better fitted for large fleets of vehicles that travel extensively, such as public transit buses or delivery trucks. As of 2015, natural gas accounted for 4% of transportation fuel use, a share that is expected to double by 2025. Natural gas vehicles (NGVs) are very similar to gasoline vehicles; the major difference is in fuel storage. Light-duty NGVs, and some heavy-duty vehicles like urban transit buses, use compressed natural gas (CNG). Heavy-duty vehicles can also use liquefied natural gas (LNG), which is denser but must be maintained below $-260\ ^{\circ}\mathrm{F}$ in very well-insulated tanks. The main markets for NGVs are new transit buses and corporate fleet cars that are used mainly for short trips. NGVs are more expensive than hybrid vehicles or gasoline vehicles. The main problem is fueling stations; a substantial portion is part of private company facilities and is not available to the general public. Natural gas for CNG is obtained directly from a distribution line. Stations require large, high-pressure compressors and storage tanks to fill a vehicle quickly. Alternatively, a small compressor can work overnight. Natural gas for LNG can also be taken from a gas pipeline and then liquefied on-site, but it also can be transported in liquid form to a refueling facility via tanker truck.

The main benefit of CNG[6] has been its relatively low price (about 80% that of gasoline). Also, the transport and distribution can rely on an existing infrastructure for both industrial and household uses. If natural gas were to be used for transportation instead of for electricity production, for example, North American natural gas reserves could supply about 20–25% of transportation fuel needs by 2020, but only with investment in distribution infrastructure. The main challenges to increased use of NGVs include an insufficient number of refueling stations and inconvenient on-board CNG tanks that take up most of the trunk space. Another key disadvantage is a limited range. The average range of a gasoline or diesel vehicle is 400 miles, and the range of an NGV is only 100–150 miles, depending on the NG compression, but the most important barrier for NGVs could be a public perception that using CNG as a fuel would involve carrying a dangerous "explosive" onboard a vehicle and that self-service refueling with a high-pressure gas would be too risky to offer to the general public. And in the end, CNG is still a fossil fuel; it is true that it pollutes the environment slightly less, but it still has all the weaknesses of fossil fuels.

---

[6] http://www.bedemes.com/en/advantages

- *Hydrogen* is often mentioned as the *energy source of the future*. The steps in using hydrogen as a transportation fuel consist of producing hydrogen via electrolysis of water or by extracting it from hydrocarbons (there are other methods as well), compressing or converting hydrogen into liquid form and storing it onboard a vehicle, and finally, using a fuel cell to generate electricity on demand from the hydrogen to propel a motor vehicle. Hydrogen fuel cells are more energy-efficient than gasoline and generate near-zero pollutants. Yet, hydrogen is also associated with several problems, particularly because a lot of energy can be wasted during its production, transfer, and storage. Hydrogen manufacturing requires electricity production. Besides, storage of hydrogen requires low-temperature/high-pressure storage tanks, adding weight and volume to a vehicle. This suggests that liquid hydrogen fuel would be a better alternative for ship and aircraft propulsion.
- *Hybrid vehicles* consist of a propulsion system using an internal combustion engine supplemented by an electric motor and batteries, which provides opportunities combining the efficiency of electricity with the long driving range of an internal combustion engine. The late 1990s saw the emergence of both, a small number of all-electric vehicles and the first gasoline hybrid vehicles. Although the all-electric vehicles did not continue in production, gasoline hybrid vehicles have continued to develop and spread in the marketplace, more recently because of higher gasoline prices and substantial tax incentives. Currently, such vehicles constitute approximately 1–2% of the light-duty vehicle fleet. Recently, there has been increased interest in developing different versions of "plug-in" hybrid electric vehicles (PHEVs) (which some are calling "extended-range electric vehicles"), although other than after-market conversions, there are more and more such vehicles on the market. There are two primary advantages that are usually cited for PHEVs. First, they will use electricity to power a portion of a vehicle's energy requirements and thus avoid some fraction of petroleum that would otherwise be consumed. This vehicle would presumably lead to reductions in petroleum imports. Second, although there would be some impact on emissions from electric power plants, vehicle emissions would be reduced especially in metropolitan and urban areas.

A hybrid vehicle still uses liquid fuel as the primary source of energy, but the engine provides the power to drive the vehicle or is used to charge the battery via a generator. Alternatively, the propulsion can be provided by the electricity generated by the battery. When the battery is discharged, the engine starts automatically without intervention from the driver. The generator can also be fed by using the braking energy to recharge the battery. Such a propulsion design significantly contributes to overall fuel efficiency, particularly in urban areas where a vehicle accelerates and brakes frequently. In the medium term, the successful development and commercialization of hybrid vehicles appear as the most sustainable alternative to conventional gasoline engine-powered vehicles. It still needs to be considered that while the emissions and pollution may be lesser (especially in urban areas), two fundamental problems remain: the first is global pollution due to electricity production, and the second is vehicle manufacture, maintenance, and disposal. The energy spent on manufacturing and recycling vehicles is a direct function of vehicle complexity, material used, fleet size, and vehicle life cycle. Hybrid vehicles actually rely on two engine systems. From this perspective, therefore, this solution is the least sustainable.

The diffusion of non-fossil fuels in the transportation sector has *serious limitations*. While the price of oil has increased over time, it has been subject to significant fluctuations. The comparative costs of alternative energy sources to fossil fuels are higher in the transportation sector than in other types of economic activities. This suggests higher competitive advantages for the industrial, household, commercial, electricity, and heat sectors to shift away from oil and to rely on solar, wind, or hydropower. Transportation fuels based on renewable energy sources might not be competitive with petroleum fuels unless there are significant price increases as well as substantial technological improvements.

- *Electricity* is being considered as an alternative to petroleum fuels as an energy source. A pure battery electric vehicle is considered a more efficient alternative to a vehicle propelled by hydrogen fuel. There is no need to convert energy into electricity because the electricity stored in the battery can power the electric motor. Besides, an all-electric car is easier and cheaper to manufacture than a comparable fuel-cell vehicle. The main barrier to the development of electric vehicles is the lack of storage systems capable of providing driving ranges and speed similar to those of conventional vehicles. The low energy capacity of batteries makes the electric car less competitive than internal combustion engines that use gasoline. The current technological level of electric cars has a range of around 400–500 km, which is steadily increasing. As technology improves, the energy and cost-effectiveness of batteries are getting better. For instance, between 2010 and 2015, the cost of lithium-ion batteries fell by 65% and still decreasing. Electric vehicles are eminently suitable for urban transportation for both passenger and freight because of the lower ranges involved. However, we need to consider that electricity is a converted energy; it is not a primary source of energy, such as fossil fuels. Moreover, this is a translation of a local problem onto a global level: the use of electric vehicles reduces local emissions but increases global pollution precisely because of the energy sources (electricity) required for their production and operation. And what exactly does the increased consumption and hence production of electricity mean is still an unexplored phenomenon! The entire issue should be addressed on a systemic level, but this remains an overwhelming, intractable problem of human understanding.

4.6.2 Air Transport/Transportation

Air transportation plays an integral part in the globalization of transportation networks. The aviation industry accounts for 8% of the energy consumed by transportation. Air transport has high energy consumption levels linked to high speeds. Fuel is the second most significant cost for the air transport industry accounting for 13–20% of total expenses. Technological innovations, such as more efficient engines and better aerodynamics, have led to a continuous improvement of the energy efficiency of each new generation of aircraft. Further distinctions in the energy consumption of transport can be made between the mobility of passenger and freight. Like other emissions resulting from fossil fuel combustion, aircraft engines, too, produce noise, gases, and particulates. Most aircraft engines are either piston engines or gas turbines, although a few have been rocket-powered, and in recent years, many small planes have used electric motors.

All except electric motors produce gases, noise, and particulates, raising environmental concerns over their global impact on air quality and global pollution. Jet planes contribute to climate change by emitting carbon dioxide (CO_2), the best-known greenhouse gas, and, with less scientific understanding, nitrogen oxides, contrails,[7] and particulates. Their radiative forcing is estimated at 1.3–1.4 that of CO_2 alone, excluding induced cirrus cloud with a very low level of scientific understanding. For example, in 2018, global commercial operations generated 2.4% of all CO_2 emissions. Jet planes have become 70% more fuel efficient between 1967 and 2007, and CO_2 emissions per Revenue Ton-kilometer (RTK) in 2018 were 47% of those in 1990. In 2018, CO_2 emissions averaged 88 g of CO_2 per revenue passenger per km. While the aviation industry is more fuel efficient, overall emissions have risen as the volume of air travel has decreased in 2019/2020 due to the pandemic but increased at the end of 2020. By the end of 2021, aviation emissions could be 70% higher than in 2005, and they could grow according to pessimistic scenario by 300% by 2050.

Aviation's environmental impact can be reduced by better fuel economy in aircraft or Air Traffic Control, and flight routes can be optimized to lower non-CO_2 impact on climate from NO_x, particulates, or contrails. Aviation biofuel, emission trading, and carbon offsetting, part of the International Civil Aviation Organization CORSIA,[8] can lower CO_2 emissions. Aviation usage can be lowered by short-haul flight bans, train connections, personal choices, and aviation taxation and subsidies. Fuel-powered aircraft may be replaced by hybrid electric aircraft and electric aircraft by hydrogen-powered aircraft. However, these are all long-term goals for which realization will be possible only after 2050.

[7] *Contrails* for "*con*densation *trails*" or *vapor trails* are line-shaped clouds produced by aircraft engine exhaust or changes in air pressure, typically at aircraft-cruising altitudes several thousand meters above the earth's surface. Contrails are composed primarily of water, in the form of ice crystals. The combination of water vapor in aircraft engine exhaust and the low ambient temperatures that exist at high altitudes allows the formation of the trails.

[8] The Carbon Offsetting and Reduction Scheme for International Aviation (CORSIA) is a carbon offset and carbon reduction scheme to lower CO_2 emissions on international flights and to decrease the aviation impact on climate change. It was developed by the International Civil Aviation Organization (ICAO) and adopted in October 2016. Its goal is to have a carbon neutral growth from 2020; what we see today is that this is an illusory goal.

4.7 The Impact of Air Transportation on Climate Change

Flights[9] are energy-intensive and depend on fossil fuels. Subsidies from fuel taxes give the airline industry an unfair advantage over other transportation modes. Consumers do not see the true environmental costs of their air travel because low flight prices do not reflect their environmental impact. Emissions from flights stay in the atmosphere and will warm it for several centuries. Because aircraft emissions are released high in the atmosphere, they have a potent climate impact, triggering chemical reactions and atmospheric effects that heat the planet. Airplanes emit gases (carbon dioxide, water vapor, nitrogen oxides, or carbon monoxide – bonding with oxygen to become CO_2 upon release) and atmospheric particulates (incompletely burned hydrocarbons, sulfur oxides, black carbon), interacting among themselves and with the atmosphere. While the main greenhouse gas emission from powered aircraft is CO_2, jet airliners contribute to climate change in four ways as they fly in the tropopause.[10] These four ways are as follows:

1. *Carbon dioxide (CO_2)*

CO_2 emissions are the most significant and best-understood contribution to climate change. The effects of CO_2 emissions are similar regardless of altitude. Airport ground vehicles, those used by passengers and staff to access airports, emissions generated by airport construction, and aircraft manufacturing and maintaining also contribute to the greenhouse gas emissions from the aviation industry.

2. *Nitrogen oxides (NO_x)*

In the tropopause, emissions of NO_x favor ozone (O_3) formation in the upper troposphere. At altitudes from 8 to 13 km (26,000–43,000 ft), NO_x emissions result in greater concentrations of O_3 than surface NO_x emissions, and these in turn have a greater global warming effect. The effects of O_3 surface concentrations are regional and local, but it becomes well mixed globally at the mid- and upper tropospheric levels. NO_x emissions also reduce ambient levels of methane, another greenhouse gas, resulting in a climate cooling effect, though not offsetting the O_3 forming effect. Aircraft sulfur and water emissions in the stratosphere tend to deplete O_3, partially offsetting the NO_x-induced O_3 increases, although these effects have not been quantified. Light aircraft and small commuter aircraft fly lower in the troposphere, not in the tropopause.

[9] https://davidsuzuki.org/what-you-can-do/air-travel-climate-change/

[10] *Tropopause* is the atmospheric boundary that demarcates the troposphere from the stratosphere, which are two of the five layers of the atmosphere of the earth. The tropopause is a thermodynamic gradient-stratification layer that marks the end of the troposphere and lies approximately 17 km (11 mi) above the equatorial regions and approximately 9 km (5.6 mi) above the polar regions.

3. *Contrails and cirrus clouds*

Fuel burning produces water vapor, which condenses at high altitude, under cold and humid conditions, into visible line clouds: condensation trails (contrails). They are thought to have a global warming effect, though less significant than CO_2 emissions. Contrails are uncommon from lower-altitude aircraft. Cirrus clouds can develop after the formation of persistent contrails and can have an additional global warming effect. Their global warming contribution is uncertain, and estimating aviation's overall contribution often excludes cirrus cloud enhancement.

4. *Particulates*

Compared with other emissions, sulfate and soot particles have a smaller direct effect: sulfate particles have a cooling effect and reflect radiation, while soot has a warming effect and absorbs heat, while the clouds' properties and formation are influenced by particles. Contrails and cirrus clouds evolving from particles may have a greater radiative forcing effect than CO_2 emissions. As soot particles are large enough to serve as condensation nuclei, they are thought to cause the most contrail formation.[11]

The contribution of jet airliners to climate change in terms of $NOx–O_3–CH_4$ interactions, contrails, the effects of soot aerosols on cirrus clouds and measuring non-CO_2 radiative forcing, has yet to be determined (Brasseur et al., 2016).

5. *Volume*

By 2018, air traffic reached 4.3 billion passengers with 37.8 million departures according to the *International Civil Aviation Organization*. The traffic was experiencing continuous growth, doubling every 15 years, despite external shocks – a 4.3% average yearly growth and forecasts expect the growth to continue. While the aviation industry is more fuel efficient, halving the amount of fuel burned per flight compared with that in 1990 through technological advancement and operations improvements, the overall emissions have risen as the volume of air travel has increased. Between 1960 and 2018, RPKs increased from 109 million to 8269 billion, and emissions increased 6.8 times (Lee et al., 2021).

Between 1990 and 2006, greenhouse gas emissions from aviation increased by 87% in the EU. In 2010, about 60% of aviation emissions came from international flights, which are outside the emission reduction targets of the *Kyoto Protocol*. International flights are not covered by the *Paris Agreement*, either, to avoid a patchwork of individual country regulations. That agreement was adopted by the *International Civil Aviation Organization*, however, capping airline carbon emissions to the year 2020 level, while allowing airlines to buy carbon credits from other industries and projects (Lowy, 2016).

[11] https://www.easa.europa.eu/eaer/air-quality/aviation-environmental-impacts

4.7.1 Evolution

By 2020, global international aviation emissions were around 70% higher than that in 2005 and they could grow by over further 300% by 2050 in the absence of additional measures.[12] The ICAO aims to lower carbon emissions through more fuel-efficient aircraft; sustainable aviation fuel; improved air traffic management; and Carbon Offsetting and Reduction Scheme for International Aviation (CORSIA). Aviation CO_2 emissions grow despite efficiency innovations to aircraft, powerplants, and flight operations. Air travel continues to increase. Many countries have pledged emission reductions for the Paris Agreement, but the sum of these efforts and pledges remains insufficient, and not addressing airplane pollution would be a failure despite technological and operational advancements. The International Energy Agency projects that the aviation share of global CO_2 emissions may grow from 2.5% in 2019 to 3.5% by 2030.[13] To the year 2020, global international aviation emissions are around 70% higher than in 2005, and the ICAO forecasts they could grow by over further 300% by 2050 in the absence of additional measures. By 2050, aviation climate impacts could be decreased by a 2% increase in fuel efficiency and a decrease in NO_x emissions due to advanced aircraft technologies, operational procedures, and renewable alternative fuels decreasing radiative forcing due to sulfate aerosol and black carbon.

4.7.2 Climate Aviation Facts

- One return flight from Canada to UK emits as much carbon emissions as heating a European home for an entire year.
- The aviation sector is responsible for 12% of transportation emissions.
- The global tourism industry is responsible for 8% of global emissions – more than the construction industry!
- Airline flying is going up 5% a year, but efficiency improvements have only increased by 1–2%.
- Airline emissions make up a little more than 3% of the total emissions in Canada.
- The total carbon impact of a single flight is so high that avoiding just one trip can be equivalent to going (gasoline) car-free for a year.

So, if you are planning a short weekend vacation, think twice before you grab that great flight deal. It is not so great when you think about the emissions that will continue to warm the planet for centuries.

---

[12] https://ec.europa.eu/clima/policies/transport/aviation_en

[13] https://www.iea.org/commentaries/are-aviation-biofuels-ready-for-take-off

4.7.3 Rail Transport

Rail transport is also known as train transport IEA (2019). It is a mode of transport on vehicles that run on tracks (rails or railroads). It is one of the most important, commonly used, and very cost-effective modes of commuting and goods carriage over long and short distances. Because this system runs on metal (usually steel) rails and wheels, it has an inherent benefit of lesser frictional resistance that helps attach more load in terms of wagons or carriages. This system is known as a train. Usually, trains are powered by an engine locomotive running on electricity or on diesel. Complex signaling systems are utilized if there are multiple route networks. Rail transport is also one of the fastest modes of land transport. Rail transport has emerged as one of the most dependable modes of transport in terms of safety. Trains are fast and the least affected by usual weather turbulences like rain or fog compared with other transport mechanisms. Rail transport is better organized than any other modes of transport. It has fixed routes and schedules. Its services are more certain, uniform, and regular.

Rail transport has also some constraints and limitations. One of the biggest constraints of rail transport is heavy cost. Trains need high capital to build and maintain, and the cost is magnified when a whole rail network is to be built. The cost of construction, maintenance, and overhead expenses are very high compared with other modes of transport. However, as we have already mentioned, we will be defenders of nature, and we will not take into account cost arguments in our reflections. Also, rail transport cannot provide door-to-door service as it is tied to a particular track. Intermediate loading or unloading involves greater cost, more wear and tear, and wastage of time.

Rail transport is often neglected about future transport systems. Probably this is because rail transport dates back centuries, to a time when it was used to help fuel the first industrial revolution. And yet, despite the advances of cars and airplanes, rail of all types has continued to evolve and thrive. In cities, metros and light rail offer reliable, affordable, and fast alternatives to road transport, reducing congestion as well as local pollution and greenhouse gas emission. High-speed rail provides a high-quality substitute for short-distance flights between major urban centers. Freight rails offer low-emission and low-cost linkage in freight supply chains. From an energy and emission perspective, rail is also among the most energy-efficient and lowest-emitting transport modes. Despite the traffic it carries, rail consumes only 2% of the total transport energy demand. Rail is also the transport sector that is most electrified and has the best and most realistic options for progress in this area. Three-quarters of passengers' movements and half of freight rely on electricity (IEA, 2019).

Rail transport is the most environment-friendly way to travel. The *greenhouse effect of gas emissions per kilometer* on railway transport is 80% less than cars. In some countries, less than 3% of all transport gas emissions come from trains. However, as the railways are electrified, this percentage could be reduced further. But, as with all other electrification attempts, we need to highlight here, too, the role and importance of primary electricity generation and its impact on pollution through greenhouse gasses and climate change.

4.7.4 Maritime Transport/Transportation

Maritime transportation accounts for 90% of cross-border world trade, as measured by volume. The nature of water transport and its economies of scale make it the most energy-efficient mode because it uses *only 7%* of all the energy consumed by transport activities, a figure way below its contribution to the mobility of goods. Nevertheless, fuel consumption is still an important factor in maritime shipping, which is related to ship design (hydrodynamics), ultilization level, operational speed, idle time (waiting at ports), and even weather conditions. For terminal operations, figures vary, but a container terminal usually has 70% of its energy consumption provided by fossil fuels (e.g., yard equipment) and 30% by electricity (e.g., portainers).

Reducing greenhouse gas (GHG) emissions is the key to avoiding the most catastrophic impacts of climate change. Countries have committed to reducing their GHG emissions under the Paris Agreement. Despite international shipping being excluded from the Paris Agreement, the International Maritime Organization (IMO) is developing its own strategy to reduce GHGs[2] from ships. The IMO member states will need to understand recent trends in ship activity and emissions to develop an effective strategy.

We know that ships accounted for approximately 1 billion tons of GHG emissions over the period 2007–2012 (IMO, 2015). However, we do not know how much GHG ships have emitted in recent years. Other information, including which ship classes emit the most GHG and under the jurisdiction of which flag states, should also be updated. Finally, policymakers would benefit from the most recent understanding of the drivers of shipping emissions (e.g., transport demand, ship capacity, and speed) in order to make informed decisions. By considering this information, IMO is more likely to reduce GHG emissions from international shipping in a targeted and cost-effective way. We found that emissions generally increased over this period, with increases in efficiency more than offset by increases in activity.

To summarize:

- Shipping GHG emissions are increasing despite improvements in operational efficiency for many ship classes. Increasing emissions are being driven by rising demand for shipping and the associated consumption of fossil fuels.
- Emissions are concentrated in a handful of ship classes and flag states. Just three ship classes (container ships, bulk carriers, and oil tankers) account for 55% of CO_2 emissions. Similarly, six flag states (Panama, China, Liberia, Marshall Islands, Singapore, and Malta) account for 52% of CO_2 emissions.
- Black carbon is a major contributor to shipping's climate impacts. On a 20-year timescale, black carbon accounts for 21% of $CO_{2\text{-eq}}$ emissions from ships.
- The biggest ships are speeding up and emitting more GHGs. Unlike most ships, the largest container and oil tankers sped up between 2000 and 2020 and became less efficient, emitting more CO_2 in 2020 than they did in 2000. As more ships follow their lead, the shipping efficiency will drop, and ship emissions will continue to rise.

- Absolute reductions in ship emissions will require concerted action to improve the energy efficiency of shipping and to develop and deploy alternative fuel and propulsion concepts. The only way to reduce emissions from ships without constraining demand is to substantially reduce the amount of CO_2 and $CO_{2\text{-eq}}$ emitted per unit of transport supply.

5 Transportation and the Environmental System

This section deals with transportation as a facet of the environmental system. The environmental dimensions of the transportation sector add up to 22% of pollution, out of which 40% are cars, 34% trucks, 11% planes, 11% ships, and 4% railway.[14]

- *Causes*. Two major factors are contributing to the level of transport activities. Economics refers to the general level of development and income. An advanced economy is likely to generate more activities per capita than a developing economy. Land use refers to the spatial structure and location of the transport demand, which indirectly influences travel distance and the modes used to support spatial interactions generated by economic activities.
- *Activities*. These involve a wide array of factors expressing the usage of transportation infrastructures and all the related services to support the transport system, particularly operations. All these activities have environmental outputs.
- *Outputs*. The first outcome of transportation activities are emissions of all sorts (carbon monoxide, nitrogen oxides, particulates, etc.). According to the geographical characteristics of the area where emissions occur (e.g., wind patterns), ambient pollution levels are created. Once these levels are correlated with population density and activity levels, a level of exposure to harmful pollutants can be calculated. This exposure is likely to have consequences.
- *End results. They include all the health, environmental, and welfare effects of the exposure to emissions from transportation activities, which are very difficult to measure* (Fig. 7.2).

The issue of transportation and the environment is *paradoxical* because transportation conveys substantial socioeconomic benefits, but at the same time, transportation is impacting the whole *environmental systems* (see also Fig. 2.11). From one side, transportation activities support increasing mobility demands for passengers and freight, while on the other, transport activities are associated with *environmental impacts and environmental footprint*. Further, environmental conditions impact transportation systems in terms of operating conditions and infrastructure requirements, such as construction and maintenance.

[14] https://transportgeography.org/contents/chapter4/transportation-and-environment/

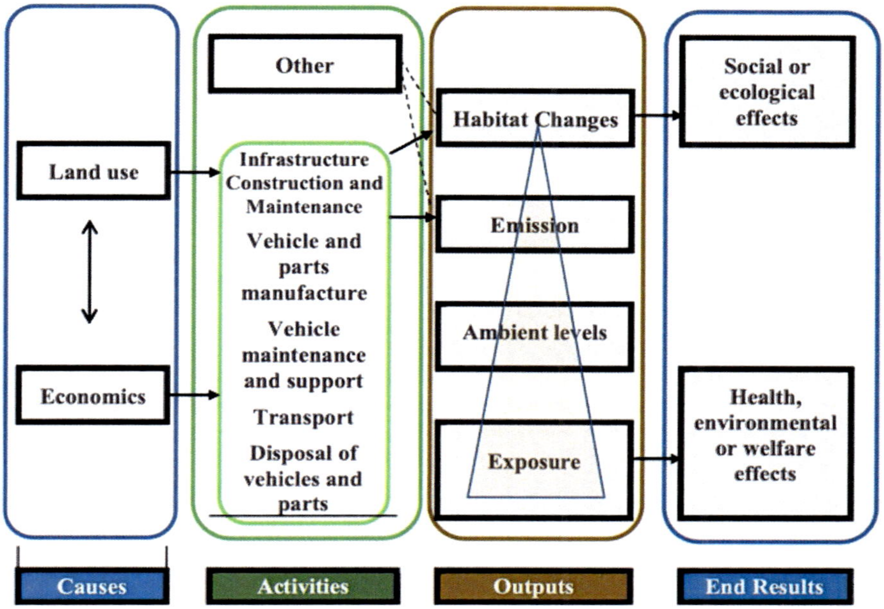

Fig. 7.2 Environmental dimensions of transportation. (https://transportgeography.org/contents/chapter3/transportation-and-society/transportation-environmental-dimensions/)

The growth of passenger and freight mobility has expanded the role of transportation as a source of emission of pollutants and their multiple *impacts on the environment*. These impacts fall within three categories:

- *Direct impacts*. The immediate consequence of transport activities on the environment where the cause and effect relationship are generally clear and well understood. For instance, noise and carbon monoxide emissions are known to have direct harmful effects.
- *Indirect impacts*. The secondary (or tertiary) effects of transport activities on environmental systems. They are often of a higher consequence than direct impacts, but the involved relationships are often misunderstood and more challenging to establish. For instance, particulates, which are mostly the outcome of incomplete combustion in an internal combustion engine, are indirectly linked with respiratory and cardiovascular problems because they contribute, among other factors, to such conditions.
- *Cumulative impacts*. They consider the varied effects on ecological footprint, of direct and indirect impacts on an ecosystem, which are often unpredictable. Climate change, with complex causes and consequences, is the cumulative impact of several natural and anthropogenic factors, in which transportation plays a role. The share of transportation in ecological footprint and global CO_2

emissions is increasing. 22% of the global CO_2 emissions are attributed to the transport sector; with this share, around 25% is for advanced economies such as the United States and even more for China's economy.

The complexities of the impacts have led to much *controversy* in environmental policy, the role of transportation, and mitigation strategies. This is made even more complex by the fact that priorities between environmental and economic considerations shift in time, which can have an impact on public policy. Public preferences concerning the environment have shifted in the last 40 years. Although environmental matters were high in the public agenda in the 1980s and 1990s, they receded in the 2000s. There are several factors at play behind these fluctuations. First, environmental concerns tend to shift with economic conditions. When economic prospects are good, environmental concerns are more prevalent, whereas when economic prospects are more limited, they are less prevalent. For instance, the recession of 2001 and the financial crisis of 2009 and 2015 were linked with sharp drops in environmental concerns (see also Fig. 2.12).

Another factor that can explain the decline in environmental concerns relates to the regulation and mitigation of the most visible environmental issues in the 1970s and 1980s, which resulted in improvements. This is particularly the case for air emission standards, such as vehicles that became less polluting and more energy efficient. Many institutions were created to regulate environmental matters for the sake of the public. Since 2010, and especially after VW diesel gate[15] in 2015, environmental concerns have been on the rise, mostly attributed to a growing awareness about climate change and its potential impacts on human activities.

The transportation sector is often subsidized, especially through the construction and maintenance of road infrastructure, which tends to be free of access. Sometimes, public stakes in transport modes, terminals, and infrastructure can be at odds with environmental issues. If the owner and the regulator are the same (different branches of the government), then there is a risk that regulations will not be effectively complied to. The total costs incurred by transportation activities, notably environmental damage, are generally not fully assumed by the users. The lack of consideration of the *real costs of transportation* could explain several environmental problems. Yet, a complex *hierarchy of costs* is involved, ranging from internal (mostly operations), compliance (abiding by regulations), contingent (risk of an event such as a spill), to external (assumed by the society).

[15]The *Volkswagen emissions scandal*, also known as *Dieselgate* or *Emissionsgate*, began in September 2015, when the United States Environmental Protection Agency (EPA) issued a notice of violation of the Clean Air Act to German automaker Volkswagen Group. The agency had found that Volkswagen had intentionally programmed turbocharged direct injection (TDI) diesel engines to activate their emission controls only during laboratory emissions testing.

5.1 Environmental Costs Hierarchy for Transportation

Cost type	Implications
Internal	Material, labor, other expenses, and revenues that are commonly allocated to a product or process. Can easily be quantified (internalized).
Compliance	Expenses incurred by and benefits to the firm that are not related to products or processes. Mostly concern compliance to regulations.
Contingent	Potential liability or benefit that depends on the occurrence of a future event. Assessed as a risk.
Image/ relationship	Costs/benefits related to the subjective perceptions of a firm's stakeholders.
External	Costs/benefits of a company's impacts on the environment and society that do not directly accrue to the business. Difficult to quantify (externalized).

For instance, external costs account on average for more than 30% of the estimated automobile ownership and operating costs. If environmental costs are not included in this appraisal, the usage of the car is consequently subsidized by society, and costs accumulate as environmental pollution. This requires due consideration as the number of vehicles, especially automobiles, is steadily increasing.

5.2 The Transport – Environment Link

The relationships between transport and the environment are a *multidimensional system*. Some aspects are unknown, and some new findings may lead to changes in environmental policies. Historically, transportation was associated with very few negative environmental impacts because of the modes used and the low mobility levels. For instance, the construction of large navies composed of sail ships was responsible for a level of deforestation in Western Europe and North America from the sixteenth to the nineteenth centuries. Urbanization in the nineteenth century and the reliance on horses created problems concerning the disposal of manure. Further, industrialization and the development of steam engines lead to pollution (e.g., soot) near ports and rail yards. Still, these issues remained marginal and localized.

However, it is only in the twentieth century that a comprehensive perspective about the links between transportation and the environment emerged, particularly with the massive diffusion of transportation modes, such as the automobile and the airplane. At the same time, manufacturing and marketing concepts such as planned obsolescence incited the design of modes such as the automobile and products (that are transported) that can continuously be replaced. The 1960s and 1970s were crucial decades in the realization of the negative environmental impacts of human activities and the need for regulations.

From an infrastructure perspective, the first comprehensive environmental regulation was established in the USA, the *National Environmental Policy Act* (NEPA), which was set in 1970 and required all federal agencies of the US government to make environmental impact assessments of their actions. Because an agency is an important provider and manager of transportation infrastructure, this legislation had substantial impacts on how transportation is assessed to be linked with environmental issues. One clear consequence was the growth in the length and the complexity of approving transport infrastructure projects to ensure they meet environmental standards. Opponents of a project could also use the regulatory framework to delay or even cancel its construction and, on occasion, change its design parameters (e.g., size). An unintended consequence was that the complexity of environmental regulations tends to impair innovations and incite current providers to keep existing infrastructure and facilities for the concern to trigger an uncertain environmental review with a new project. In time, this slowed down the development of transport infrastructure and substantially increased their costs.

From an operational perspective, the Clean Air Act of 1970 set clear air quality standards and expectations for both stationary (e.g., a power plant) and mobile (e.g., an automobile) sources of air pollutants. The US Environmental Protection Agency (EPA)[16] first approved a waiver of the Clean Air Act to allow the standards to move forward. In January 2009, however, EPA began a formal process to reconsider the waiver. Finally, in May 2009, after detailed discussions among California, EPA, and auto makers, an approval of the waiver was announced, as well as an approval of a new unified approach to both federal corporate average fuel efficiency (CAFE)[17] and GHG emissions standards, as the basis for a national standard. This action is expected to result in the achievement of the former 35.5 miles per gallon CAFE goal by 2016, several years sooner than originally envisioned. For transportation, it immediately set emissions standards for a list of acknowledged pollutants, such as carbon dioxide, volatile organic compounds, and nitrogen oxide. The outcome was a rapid decline in air pollutant emissions by the transportation sector through better engine technology. The Clean Water Act of 1977 provided a similar regulatory environment concerning water pollution and the ability to build infrastructures over wetlands.

The 1990s was characterized by a realization of global environmental issues, epitomized by the growing concerns between anthropogenic effects and climate change. Transportation also became an important dimension of the concept of sustainability, which has become a core focus, ranging from vehicle emissions to green supply chain management practices. These developments require a deep understanding of the reciprocal influence between the physical environment and transport infrastructures, and yet this understanding is often lacking. The main factors considered in the physical environment are geographical location,

---

[16]The US Environmental Protection Agency (EPA) is responsible for the protection of human health and the environment. EPA: Provides technical assistance to support recovery planning of public health and infrastructure, such as wastewater treatment plants.

[17]Corporate Average Fuel Economy (CAFE) standards regulate *how far our vehicles must travel on a gallon of fuel*. CAFE standards set fuel consumption for passenger cars and for light trucks (collectively, light-duty vehicles) and separately set fuel consumption standards for medium- and heavy-duty trucks and engines.

topography, geological structure, climate, hydrology, soil, natural vegetation, and animal life.

The environmental dimensions of transportation are related to the *causes*, *activities*, *outputs*, and *results* of transport systems. Establishing linkages between environmental dimensions is a difficult undertaking. For instance, to what extent carbon dioxide emissions are linked to land use patterns? Furthermore, transportation is embedded in environmental cycles, notably over the carbon cycle where carbon flows from one element of the biosphere, like the atmosphere, to another like the ecosphere, where it can be accumulated (permanently or temporarily) or passed on. The relationships between transport and the environment are also complicated by two observations:

- *Level of contribution*. Transport activities contribute among other anthropogenic and natural causes, *directly, indirectly, and cumulatively to environmental problems*. In some cases, they may be a dominant factor, while in others, their role is marginal and challenging to establish.
- *Scale of impact*. Transport activities contribute at different geographical scales to environmental problems, ranging from local (noise, particles, CO, CO_2, NO_x emissions, etc.) to global (climate change), not forgetting continental/national/regional problems (smog and acid rain).

Establishing environmental policies for transportation thus must take account of the level of contribution and the geographical scale; otherwise, some policies may just move the problems elsewhere and have unintended consequences. Environmental policies in advanced economies could incite the relocation of some activities with high environmental externalities (e.g., steel making) in developing countries. This transfers the problem from one location to another. Still, such a transfer usually involves new equipment and technologies that are usually having a lower environmental impact. Even if an administrative division (municipality, county, state) has adequate environmental enforcement policies, the geographical scale of an environmental impact (notably air, water, and soil pollutants) goes beyond established jurisdictions. This has become salient in the disposal of waste such as electronic goods that are transferred to developing economies with lower environmental regulations to be disposed of or recycled.

The structure of the transport network, modes used, and traffic levels are the main factors of the environmental impact of transportation. Networks influence the spatial distribution of emissions, while modes relate to the nature of the emissions and the traffic to the intensity of these emissions. In addition to these environmental impacts, technological processes sustaining the transport system must be considered. These include the extraction and production of fuels, vehicles, and construction materials, some of which are very energy-intensive (e.g., aluminum and steel), and the disposal of vehicles and parts as well as the provision of infrastructure. They all have a life cycle timing their production, utilization, and disposal. Thus, the evaluation of the link between transport and the environment without the consideration of cycles in the environment and in the product, life alike, is likely to convey a limited overview of the situation and may even lead to incorrect appraisal, policies, and mitigation strategies.

5.3 *Environmental Dimensions*

Transportation activities support increasing mobility demands for passengers and freight, notably in urban areas. But transport activities have resulted in growing levels of motorization and congestion. As a result, the transportation sector is becoming increasingly linked to environmental problems.

6 Climate Change

The greenhouse effect is a component of the global climate and is a naturally occurring process that involves global warming. Greenhouse gases include carbon dioxide (CO_2), methane (CH_4), nitrous oxide (N_2O), and halocarbons, gases that accumulate in the atmosphere long enough to reach a homogeneous composition across the world. Thus, irrespective of the location, their concentration is similar. The quantity of conventional greenhouse gases released into the atmosphere has increased substantially since the industrial revolution and particularly over the last 35 years. The respective impacts of greenhouse gases are further complex and complicated by differences in their atmospheric lifetime (or residence time), which is the time they spend in the atmosphere before decaying or being absorbed by biological or chemical processes. For CO_2, it can range between 5 and 200 years, while it is about 12 years for methane and 114 years for N_2O. For halocarbons, such as chlorofluorocarbons, it is at least 45 years.

The activities of the transport industry release several million tons of greenhouse gases each year into the atmosphere, accounting between 25% and 30% of all greenhouse gas emissions. There is an ongoing debate about to what extent these emissions are linked with climate change, but the debate relates more to the extent of these impacts than their nature. Some gases, particularly nitrogen oxide, also participate in depleting the stratospheric ozone (O_3) layer, which naturally screens the earth's surface from cosmic radiation. The rise in air traffic, in addition to its emissions, has increased the number of contrails, which are mainly ice crystals formed from condensation around planes flying at high altitudes. They can contribute to climate change in a paradoxical fashion as, on the one hand, they can trap heat and, on the other, also reflect solar radiation. In addition to being a contributor to climate change, transportation is also impacted by it, particularly over infrastructure (e.g., more floods due to rising sea levels) and operations (harsher operating conditions) (Fig. 7.3).

	Operations	Infrastructure
Heat waves	• Impact of lift-load limits on shorter runways • Limits on periods of construction activity	• Thermal expansion of bridges. • Pavement integrity and softening • Deformation of rail tracks.
Rising see levels	• Frequent interruptions of costal low-lying road, rail and air transport	• More frequent flooding of infrastructure (and potential damage) in low lying areas • Erosion of infrastructure support. • Changes in harbor facilities to accommodate higher. tides and surges.
Intensity of precipitation	• Increase in weather related delays and disruption, particularly road and air transport	
More frequent hurricanes	• Frequent interruption of services • Frequent and extensive evacuations of coastal areas • Debris of road and rail infrastructure.	• Greater probability of infrastructure failure. • Greater damage to port infrastructure
Increase in arctic temperature	• Longer shipping seasons • More ice-free ports in northern regions. • Availability of trans-arctic shipping routes.	• Damage to infrastructure. because of the thawing of permafrost. • Shorter season for ice-roads.

Fig. 7.3 Climate changes and its potential impacts on transportation

6.1 Transportation, Sustainability, and Decarbonization

Sustainable transportation is the capacity to support the mobility needs of a society in a manner that is the least damageable to the environment and does not impair the mobility needs of future generations.

6.1.1 Sustainable Transportation

Transportation, as a core component supporting the interactions and the development of socioeconomic systems, has also been the object of much consideration to what extent it is sustainable (Black, 2010; Schiller & Kenworthy, 2018; Tolley, 2003). Sustainable transportation is the capacity to support the mobility needs of a society in a manner that is the least damageable to the environment and does not impair the mobility needs of future generations.[18]

Sustainable development[19] applied to transport systems requires the promotion of links between environmental protection, economic efficiency, and social progress (Banister & Button, 1993; Banister, 2008). Under the environmental dimension, the objective includes understanding the reciprocal influences of the physical environment and the practices of the industry and that environmental issues are addressed by

[18] https://transportgeography.org/contents/chapter4/transportation-and-environment/

[19] https://transportgeography.org/contents/chapter4/transportation-sustainability-decarbonization/sustainable-transportation/

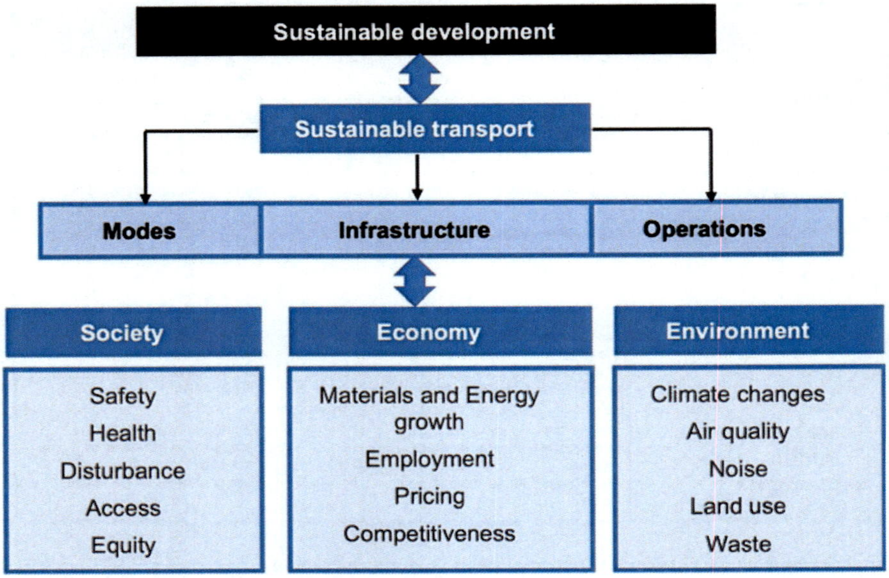

Fig. 7.4 Sustainable transportation

all aspects of the transport industry. Under the economic dimension, the objective is to achieve progress in terms of economic efficiency. Transport must be cost-effective and capable of adapting to changing demands (Hickman et al., 2013). Under the social dimension, the objective consists in upgrading standards of living and quality of life, which is symbolically shown in Fig. 7.4.

While policies, rules, and regulations expect compliance, users tend to instinctively react to price signals and discard modes that are becoming costly (unsustainable) and find loopholes. Transportation and sustainability for both passengers and freight must also contend with mitigation versus adaptation issues:

- *Mitigation* concerns the improvement of productivity and efficiency of existing modes, terminals, and managerial approaches so that environmental externalities are reduced. They tend to be short- to medium-term strategies.
- *Adaptation* is a change in the level of use and the market share of respective modes to reflect better a long-term trend, such as higher energy prices, improved information technologies, and stricter environmental regulations.

6.1.2 Mitigation – Vehicle Efficiency

Few aspects of using energy have such a direct impact on consumers as the efficiency of our vehicles, the vast majority of which are powered by internal-combustion engines burning gasoline or diesel fuels. Thanks to decades of scientific and engineering progress, efficiency has improved dramatically in recent decades, with more advances on the way. On average, vehicles have gotten 20% heavier since 1980. Yet in that same period, horsepower has increased more than 80% and fuel economy by 100%. In passenger cars, this has resulted in a 37% decline in energy intensity.

Modern vehicles approach the goal of increased efficiency in many ways, including increasing the compression ratio,[20] replacing carburetors with fuel injectors that pump atomized fuel into the cylinders, and doubling (generally from two to four) the number of valves per cylinder, for more efficient transport of intake and exhaust. Additional performance increases result from turbochargers (now found in more than 20% of all new cars and 14% of trucks) that thoroughly mix air and fuel via a turbine that propels the mix into the cylinder at above-atmospheric pressure. As a result, we now get sufficient power from ever-smaller engines. In 1980, the average displacement of passenger-car engines was approx. 3.0 L. By 2013, it was approx. 2.0 L, which indicates more than a 30% reduction. By 2020, the trend dropped to 1.6–1.3-L engines, indicating an additional decrease.

Gears change the ratio between how fast the engine crankshaft turns and how fast the wheels turn (travelling speed), allowing the engine to stay at an efficient number of revolutions per minute at different vehicle speeds. For years, three- to six-speed transmissions were commonplace. The newest vehicles, however, have automatic transmissions with as many as nine or more gears, which allow the engine to operate at lower revolutions per minute while delivering optimal torque at a given vehicle speed. The test results show that changing from a conventional five-speed automatic transmission to a modern eight-speed, dual-clutch transmission system can reduce total fuel consumption by 11% in compact passenger cars and 8.5% in sport utility vehicles.

The motion of a vehicle is impeded by the rolling friction between the tires and the road surface. Modern tires incorporate materials and operate at pressures that minimize the energy loss caused when a pneumatic tire deforms and sticks to the road. At the same time, vehicles are subject to air resistance (drag), which has been progressively minimized in more streamlined designs. In the case of modern passenger cars, such as sedan, the drag coefficient is about 0.3, but due to the emergence of SUV vehicles, it is growing again, which also affects fuel consumption and consequently environmental pollution.

Taken together, these innovations have resulted in substantial improvements and reduction in total fuel consumption by more than 50%.

6.1.3 Adaptation – CAFE Standards

One of the most impressive efficiency successes in modern memory is the result of the federal Corporate Average Fuel Economy (CAFE) standards established in 1975. They stipulated that the average fuel economy for new cars, sport utility vehicles

[20] *Compression ratio* is how much the air–gas mixture is compressed in the cylinder before ignition.

(SUVs), and light trucks would be 25 miles per gallon (mpg) by model year 1985. They stipulated that the average fuel economy for new cars, sport utility vehicles (SUVs), and light trucks would be 25 miles per gallon (mpg) by the year 1985 – which is a nominal improvement of nearly 50% in comparison to the stipulated 18 mpg for the year 1978. (CAFE standards are calculated by a production-weighted average of mileage ratings across a manufacturer's fleet; they are not based on data from actual on-road performance.) The US Department of Transportation specified an average for light trucks of 20.7 mpg. Automakers complied, dramatically improving the fuel economy of the nation's light-duty vehicle fleet, reducing dependence on imported oil, improving the nation's balance of trade, and reducing carbon dioxide (CO_2) emissions. In December 2007, the Congress passed an updated CAFE law mandating that new cars, SUVs, and light trucks together average 35 mpg by 2020, an increase of 40% from the previous 25-mpg average. By 2014, manufacturers were meeting standards of 34.2 mpg for passenger cars and 26.2 mpg for light trucks. *The most recent federal efficiency standards are projected to increase fuel economy to the equivalent of 54.5 mpg for cars and light-duty trucks by model year 2025 while also reducing CO_2 emissions.*[21]

7 Push for Decarbonization

Decarbonizing transportation aims to reduce, mitigate, and even eliminate carbon emissions by adapting transportation infrastructures, conveyances, and operations.

The concept of sustainable transportation has become widely accepted as a goal and appears in the environmental plans of many governments and corporations. Still, sustainable transportation remains elusive as it does not offer clear guidelines but mostly a narrative allowing stakeholders to remain vague in their commitments and endeavors. For instance, in 2015, the United Nations articulated 17 sustainable development goals covering an outstanding range of issues ranging from reducing economic and gender inequality to eradicating poverty and hunger. Transportation was not identified as a distinct sustainability goal, even if it accounts for about a quarter of global CO_2 emissions. Further, these goals have no stated priority and are, therefore, subject to interpretation (McKinnon et al., 2013; McKinnon, 2018a, b).

Since the early 2000s, a reframing of sustainability goals in the transportation sector took place toward a more tangible strategy focusing on carbon. This mainly took shape around the decarbonization of transportation, which helps articulate the narrative around fossil fuels (Noussan et al., 2020). The concept does not undermine the purpose of transportation, which is providing mobility to passengers and freight, but that the carbon footprint of transportation activities should be reduced. Even if it focuses on carbon, decarbonization directly impacts other externalities as most air

[21] https://www.eia.gov/energyexplained/use-of-energy/transportation.php

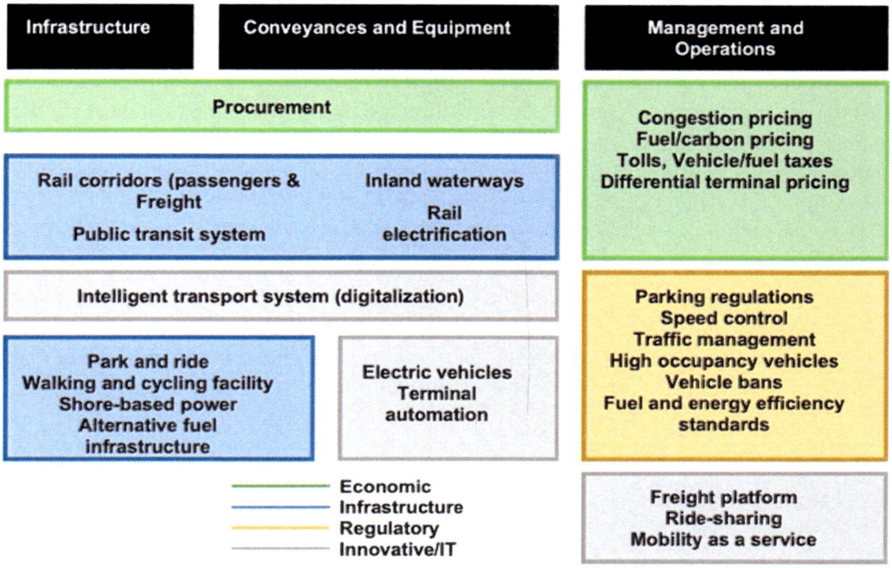

Fig. 7.5 The decarbonization of transportation

pollutants are an outcome of the combustion of fossil fuels. Decarbonizing transportation focuses on three main realms of application (see Fig. 7.5):

- *Infrastructure*. The fixed-asset components of decarbonization include transport corridors and terminals. Their construction, maintenance, and upgrade can be subject to procurement strategies that are less carbon-intensive, including the use of materials. Transportation modes, particularly in terms of the economies of scale they offer, can be ranked by their carbon intensity. This implies that infrastructure-supporting modes with low carbon intensity should be favored as well as the connectivity between transportation modes.
- *Conveyances and equipment*. For mobile transportation assets, the focus is mainly on their fuel and sources of energy. The electrification of roads and rail is seen as a key strategy as it focuses on modes that have the highest contribution to CO_2 emissions. Ideally, urban mobility should have a larger share accounted for by pedestrians and bicycles, but humans also need energy and food, which must be produced (see Fig. 7.3 – and *Level of Contribution and Scale of Impact*).
- *Management and operations*. A focus is on pricing strategies that change the competitiveness of transportation modes according to their carbon emission. The expectation is that the increasing competitiveness of decarbonized transportation will displace transportation technologies based on fossil fuels. Better utilization of existing transportation assets is also recognized, such as freight platforms and ride-sharing services.

The push toward the decarbonization of transportation will continue until concerns about climate change have abated. This would imply that the environmental impacts of human activities have been mitigated to a level that is considered to be sustainable. Sustainability and decarbonization are part of the same agenda, with the main difference being that decarbonization offers a clearer framework and plan to action with practical solutions. But decarbonization is only a small part of the greater environmental issue associated with transportation.

As Dieselgate or Emissionsgate have shown, there are a number of other pollutants out there, from particulates to NOx nitrogen oxides, which are just as harmful, or more, than carbon dioxide itself. It is no longer just about the carbon footprint; we need to start talking about the environmental footprint, which looks at the problem of pollution in a complex and comprehensive manner, rather than just partially, in terms of decarbonization, carbon footprint, or climate change!

8 Health and GHM and Other Non-GHG Damages

The attempt to estimate the health and non-GHG damages on a per-gallon (liter) basis is complicated by the fact that simply multiplying expected miles per gallon for each fuel and vehicle type by the damages per mile tends to make the most fuel-efficient vehicles appear to have higher damages per gallon than less fuel-efficient vehicles. It is important in interpreting such results to consider major limitations in the analysis:

- *Emissions and damages are not quantifiable.* Although analyses were able to consider and quantify a wide range of emissions and damages throughout the life cycle and included the most significant contributors to estimates of such damages, there are many potential damages that could not be quantified. Such damages include the following:
  - *Overall:* Impacts of hazardous air pollutants and damages to ecosystems (e.g., from deposition), the full range of agricultural crops, and others.
  - *Biofuels:* Impacts on water use and water contamination, as well as any formal consideration of potential indirect land-use effects.
  - *Battery electric vehicles:* Potential effects from exposures to air toxics in battery manufacture, in battery disposal, and during accidents.
  - *Uncertainty.* Any such analysis includes a wide set of assumptions and decisions about analytical techniques that can introduce uncertainty in the results.

8.1 Overall Implications of the Results

Perhaps the most important conclusion to be taken from different analyses of *GHG Damages* is that when viewed from a full life-cycle perspective, the results are remarkably similar across fuel and technology combinations. One key factor contributing to this result is the relatively high contribution of emissions to health and other non-GHG damages in life-cycle phases (such as those in the development of the feedstock, processing of the fuel, and manufacturing of the vehicle) other than in the phase of vehicle operation.

There are some differences though, and from these, some conclusions can be drawn:

- The fossil fuel-driven technologies have somewhat higher damages and GHG emissions in 2020 than several other fuel and technology combinations. The grid-dependent electric options have somewhat higher damages and GWP than other technologies in large measure due to the continued conventional and GHG emissions from the existing and likely future grid at least as of 2030.
- In 2030, with the move to meet the enhanced 35-mpg requirements now being put in place, those differences among technologies tend to converge somewhat, although the fact that operation of the vehicle is generally less than a third of overall life-cycle emissions and damages tends to dampen the magnitude of that improvement. Further enhancements in fuel efficiency, the likely push for an extension beyond 2020 to further improvements, would improve the GHG emission estimates for all liquid–fuel-driven technologies.
- The choice of feedstock for biofuels can significantly affect the relative level of life-cycle damages, with herbaceous and corn stover having some advantage in this analysis.
- Additional regulatory actions or changes in the mix of electricity generation can significantly affect the levels of damages and GHG emissions. Similarly, major regulatory initiatives to reduce electricity-generation emissions or legislation to reduce carbon emissions would significantly improve the relative damages and emissions from the grid-dependent electric options. A shift to electricity generation with lower emissions (e.g., natural gas, renewables, and nuclear) would also further reduce the life-cycle emissions and damages of the grid-dependent technologies.
- Overall, the differences are somewhat modest among different types of vehicle technologies and fuels, even under the likely 2030 scenarios, although some technologies (e.g., grid-dependent electric) had somewhat higher life-cycle emissions. Therefore, some breakthrough technologies (such as cost-efficient conversion of advanced biofuels, cost-efficient carbon capture and storage, and much greater use of renewable resources for electricity generation) appear to be needed to dramatically reduce transportation-related externalities.

Developments in automotive technology have had a profound effect on the energy sector. The electronics and computer revolutions have made possible very small sensors and integrated circuits. Many of these are used in vehicles to continuously monitor and optimize engine and drivetrain performance. Concurrently, innovations such as tires with lower rolling resistance, more efficient multi-speed transmissions, turbocharged and/or multi-valve engines that get more horsepower out of fuel, growing numbers of front-wheel-drive vehicles, and greater electrification of drivetrains such as with hybrid electric vehicles, among many other advances, have led to improved fuel efficiency and to more power, better control, and lower emissions.

However, such sustainable alternatives contrast with the reality, and most people still prefer automobiles and trucks. Thus, *sustainable transportation remains elusive* because any economic activity, including transportation, has negative environmental impacts. The matter remains if these activities are taking place at a level exceeding the environmental and social carrying capacity. Technological innovation has historically played a paradoxical role of both exacerbating environmental and sustainability issues and, at the same time, offering forms of mitigation. The expectation is that in the future, technological innovation in the transport sector will be a more sustainability driver than it was in the past. This is the main reason why a share of the attention has shifted towards decarbonizing transportation.

The impacts of transportation on the environment (and the environmental footprint) are complex and multifaceted. There is a need for additional research to attain the following:

1. At the earliest possible stage in the research and development process, better understanding of the potential negative externalities for new fuels and technologies should be acquired to avoid these externalities as the fuels and technologies are being developed.
2. Understanding of the currently unquantifiable effects and potential damages should be improved, especially as they relate to biofuels (such as effects on water resources and ecosystems) and battery technology (such as effects throughout the battery life cycle of extraction through disposal).
3. More accurate emission factors should be obtained for each stage of the fuel and vehicle life stages. In particular, there is a need, in the context of enhancing even further EPA's recent shift to the Motor Vehicle Emission Simulator (MOVES) model for mobile-source emissions, to make measurements to confirm or refute the assumption that all vehicles will only meet but not exceed the emission standards. In actual practice, there can be significant differences between on-road performance relative to emission requirements, and some alternative-fuel vehicles may do better or worse than expected.
4. The issue of indirect land-use change is central to current debates about the merit of biofuels. Regardless of whether this impact is regarded as an externality associated with world biofuel production, it is important to obtain more empirical evidence about its magnitude and causes, as well as to improve the current suite of land-use change models.
5. Because a substantial fraction of life-cycle health impacts comes from both vehicle manufacture and fuel production, it is important to improve and expand the information and databases used to construct emission factors for these life stages. In particular, there is a need to understand whether and how energy-efficiency improvements in these industrial components might change the overall estimates of life-cycle health damages.

6. As better data become available, future studies should also focus on other transportation modes, those that are alternatives to automobiles and light truck (transit), as well as air, rail, and marine, which are alternatives for long-distance travel and for freight.

8.2 Environmental Commitment

Global warming and climate change are inescapable facts that open up a wide range of political and economic issues, affecting society as a whole, the world economy, technology, and the environment. Atmospheric emissions have been rising over the past decades, and the global average temperature in 2017 approached 1.1 °C relative to pre-industrial levels. Four years later, in 2021, we have already recorded an increase of almost 2 °C, which is extreme. Moreover, this trend continues to rise despite a series of grand and noble agreements, which mostly only exist on paper.

The underlying cause of emissions is mainly the extraction of energy from fossil fuels, upon which years of economic development since the first industrial revolution have relied, and still rely today. Climate change is manifested in the form of melting glaciers, rising sea levels, extreme weather events, soil erosion, polluted atmosphere, loss of plant and animal species, and global changes in ecosystems. Consequently, the adoption and, above all, the implementation of the commitments of global climate agreements are vital to the existence of civilization. The EU, and sadly only a few other superpowers, are particularly well aware of the fact that fighting climate change requires coordinated action on behalf of all stakeholders. The EU has taken the lead in combating climate change and implementing commitments to various international agreements, such as the Kyoto and Paris Agreements, and the UN Environment Convention, whose main objectives have been to limit global warming to less than 2 °C relative to pre-industrial levels and to reduce the environmental footprint. The baseline problem here is the definition of "pre-industrial level," which is interpreted differently by different stakeholders. The EU is proof that measures to combat climate change can be effective, as greenhouse gas emissions were reduced by 23% between 1990 and 2014, while the gross domestic product increased by 46%. In the context of climate policy, the EU has set three sets of goals at the beginning of the new millennium: goals to be reached by the year 2020, goals to be reached by 2030, and finally, long-term goals to be reached by 2050, with a full transition to a low-carbon, climate-resilient society.

9 United Nations Framework Convention on Climate Change

The United Nations Framework Convention on Climate Change (UNFCCC) is an international environmental treaty adopted by the United Nations Conference on Environment and Development (UNCED), informally also known as the "Earth

Summit" held in June 1992 in Rio de Janeiro. Its central objective was to stabilize the concentrations of greenhouse gases in the atmosphere at a level that would prevent dangerous anthropogenic interference with the climate system. The treaty itself did not set mandatory GHG emission limitations for individual countries, nor did it set out mechanisms for enforcing them. Consequently, the UNFCCC is considered not to be legally binding for the signatories. However, the treaty did establish protocols for mandatory emission limit values (UNFCCC, 1992). One of the most well known is the Kyoto Protocol, which is discussed in more detail below.

The UNFCCC was signed on 9 May 1992 and entered into force on 21 March 1994; by May 2004, a total of 194 countries had signed the convention. One of the main tasks of the UNFCCC was to establish national GHG inventories, the objective of which is to determine the magnitude of national GHG emissions and removals that are directly attributable to human activity. The signatories to the convention meet each year to assess progress in addressing climate change. The design of the treaty was a framework for the participation of countries in limiting the rise in the average temperature of the planet, fighting climate change, and tackling the consequences of these changes.

The UNFCCC divided the signatories into three groups as follows (UNFCCC, 1992):

- Industrialized countries and economies in transition
- Developed countries that pay the costs of developing countries
- Developing countries

Industrialized countries and economies in transition have committed to reducing GHG emissions to 1990 levels. Producers can only exceed their allocated emissions if they buy carbon offsets, or if they correct their surpluses through a mechanism agreed by all signatory states.

The second group of countries are the developed countries of the world, members of the OECD, except for those countries that were still in transition in 1992. The third group of countries are developing countries, which do not have to reduce GHG emissions unless developed countries provide sufficient funding and appropriate technology to enable this. Developing countries may voluntarily join the group of industrialized countries if they consider that they are sufficiently developed.

Since the entry into force of the UNFCCC, the parties have met annually at the Conference of the Parties to assess the progress in tackling climate change and to set legally binding obligations for states to reduce GHG emissions (UNFCCC, 1992).

10 The Kyoto Protocol

UNFCCC signatories soon realized that stricter provisions were needed to reduce emissions, which would be legally binding. As a result, 1997 saw the signing of the Kyoto Protocol, which introduced legally binding goals for reducing GHG emissions in developed countries (European Commission, 2018a). The ratification of the

Doha Amendments under the Kyoto Protocol also sets out the objectives for the second period, from 2013 to 2020.

The Kyoto Protocol is an international agreement directly linked to the UNFCCC, which binds its parties to reduce emissions. Developed countries have acknowledged that they are mainly responsible for the high state of GHG emissions resulting from more than 150 years of industrialization (UNCC, 2018). The main goal of the Kyoto Protocol is to reduce emissions of carbon dioxide and five other greenhouse gases.

Under the Kyoto Protocol, it is also extremely important to monitor the actual emissions of individual countries and to keep accurate records of transactions. The reporting system monitors and records all transactions by the Parties under three Kyoto Protocol mechanisms. To this end, the UN Climate Change Secretariat, based in Bonn, Germany, keeps an international transaction log. Parties thus submit annual emission inventories and national reports in accordance with the Kyoto Protocol (UNCC, 2018) on a regular basis.

10.1 The Paris Agreement

The Paris Agreement is considered to be the first legally binding international climate agreement with legal and economic consequences for the signatory states. The EU played a leading role in its adoption, content design, and implementation of commitments. In the following section, we will briefly present the content of the agreement and the role played by the EU.

10.2 The Role of the EU in Drawing Up the Paris Agreement

At the beginning of the twenty-first century, the EU has taken the lead in combating climate change. As a result, the 21st United Nations Conference on Climate Change in Paris was organized under the auspices of the EU, also known as COP 21 (Conference of Parties 21), held from 30 November to 12 December 2015. At the Paris Conference on Climate Change, 195 countries adopted a new universal and first legally binding global climate agreement (European Commission, 2018b).

After more than 20 years of negotiations at the United Nations, the EU has succeeded in reaching a legally binding and universal agreement on climate change, and Europe has taken the lead in becoming the first region with binding objectives, placing it well on its way to becoming a world leader in the field of renewable energy sources (European Commission, 2018c). The EU has demonstrated that it is indeed possible to take action in the field of reducing emissions. Between 1990 and 2014, emissions in the EU was reduced by 23%, while GDP increased by 46%. The European Commission's (European Commission, 2018c) standpoint is that climate change threatens the living environment, destabilizes society, and limits further economic development. The worst consequences of climate change could be the displacement of hundreds of millions of people as a result of erosion and flooding of coastal areas and additional GHG emissions which could lead to further warming and long-term changes in the climate system. In the eyes of the EC, the time has come for decisive, global action against climate change, which is at the same time an economic opportunity for EU citizens, businesses, and EU Member States.

The climate agreement reached was the result of many years of negotiations by the international community, which sought to reach a multilateral agreement. Following the limited participation of countries in the Kyoto Protocol and the disagreements of the countries in Copenhagen in 2009, the EU began to seek and form a broad alliance and coalition of developed and developing countries, with the aim of achieving a common ambitious goal (European Commission, 2018d). The Paris Agreement gave a clear signal to investors, businesses, and policy makers that it is time to move on to clean, sustainable energy (Kutscher et al., 2018) and that fossil fuels need to be replaced by green energy sources that are environmentally friendly. The then European Commissioner for Climate Action and Energy Miguel Arias Cañete said (European Commission, 2015c, p. 1):

> *This agreement is a major win for Europe. But more importantly, it is a major win for the global community. Europe has led the efforts in Paris to get an ambitious and legally binding global climate deal. We have forged alliances and others have joined. Our key objectives – on the long-term goal, the 5-yearly review cycles and transparency – are in the new agreement. The agreement also reconfirms global commitment to continued support to those in need of assistance. We succeeded. Now, what has been promised must be delivered. Europe will continue to lead the global low-carbon transition we have agreed.*

The EU's vision for a trustworthy agreement includes the following (European Commission, 2015b):

- a global vision of the long-term goal as a message to all stakeholders, including businesses, investors, and the public, about the determination to move to a low-carbon economy;
- a mechanism to regularly review and increase common ambitions; and
- a sustainable system of transparency and accountability to ensure to stakeholders and interest groups that promises will be kept (Kutscher et al., 2018).

11 The Long-Term Goal – The Green Deal

Climate change and environmental degradation are an existential threat to Europe and the world. To overcome these challenges, the European Green Deal will transform the EU into a modern, resource-efficient, and competitive economy, ensuring:

- no net emissions of greenhouse gases by 2050,
- economic growth decoupled from resource use,
- no person and no place left behind.

The European Green Deal is also our lifeline out of the COVID-19 pandemic. One-third of the 1.8 trillion-euro investments from the Next-Generation EU Recovery Plan and the EU's 7-year budget will finance the European Green Deal. On 14 July 2021, the European Commission adopted a set of proposals to make the EU's climate, energy, transport, and taxation policies fit for reducing net greenhouse gas emissions by at least 55% by 2030, compared with the levels in 1990. Achieving these emission reductions in the next decade is crucial to Europe becoming the world's first climate-neutral continent by 2050 and making the European Green Deal a reality.

11.1 Clean Energy Transitions

The European Green Deal focuses on three key principles for the clean energy transition, which will help reduce greenhouse gas emissions and enhance the quality of life of our citizens:

1. Ensuring a *secure* and *affordable EU energy supply*
2. Developing a *fully integrated*, *interconnected*, and *digitalized EU energy market*
3. Prioritizing *energy efficiency*, improving the *energy performance of our buildings* and developing a power sector based largely on *renewable sources.*

The Commission's main objectives to achieve this are as follows:

- *Build interconnected energy systems* and better integrated grids to support renewable energy sources
- Promote innovative technologies and *modern infrastructure*
- Boost *energy efficiency* and *eco-design* of products
- *Decarbonize the gas sector* and promote *smart integration* across sectors
- *Empower consumers* and help EU countries tackle energy poverty
- Promote *EU energy standards* and technologies at the global level
- Develop the full potential of Europe's *offshore wind* energy

The European Commission adopted a set of proposals to make the EU's climate, energy, transport, and taxation policies fit for reducing net greenhouse gas emissions by at least 55% by 2030, compared with the levels in 1990. Based on these proposals, at least three issues which began to emerge at the end of 2021 can be pointed out:

- The problem of the cost of achieving the set goals
- The problem of lack of sustainable energy sources
- The problem of globalization (we must be aware that the EU is only a small part of the global problem, with a limited impact on the proposed solutions)

But let us return once again from the level of local (EU) problems and EU commitments to the global situation. The recently held 2021 United Nations Climate Change Conference COP26 (UNCC_COP26, 2021) was a chance for leaders from all countries in the world to make the year 2021 the necessary turning point for taking action on the climate emergency (UNCC_COP26, 2021).

12 "Blah, Blah, Blah"

Like so many times before, the story (COP26) is repeated, grand words at the beginning and hesitant (non-) conclusions at the end. We already know about everything that has been said. It has all been told over and over again. Most of us agree that the problem exists, but no one is willing to act. As the old phrase says,

The dogs bark, but the caravan goes on!

However, unlike previous false promises made by influential politicians and the world's most powerful leaders, a tiny but hopeful light did shine at the end of this tunnel of fruitless negotiations. This light was lit by a wave of youngsters, unspoiled by capital and lies, who began to realize the seriousness of the situation. On behalf of these youngsters, Greta Thunberg criticized leaders over the climate crisis, saying:

They invite cherry-picked young people to meetings like this to pretend that they listen to us. But they clearly don't listen to us. Our emissions are still rising. The science doesn't lie.

We can no longer let the people in power decide what is politically possible. We can no longer let the people in power decide what hope is. Hope is not passive. Hope is not blah, blah, blah. Hope is telling the truth. Hope is taking action. And hope always comes from the people.[22]

Many words have been said and written down, but most remained only words on paper, or not even that. People in power should just remember the question: *Every vote counts! But, do all votes count the same?* And the answer: *Certainly, the voice of the everyman does not count as much as the voice of an influencer.* In the Forbes magazine analyses, the starting point is the everyman:

1 vote of the everyman counts 1 vote
1 vote of the most influential politician counts approx. 790 million votes
1 vote of one of the world's most powerful people counts 105 million votes
1 vote of a billionaire counts approx. 2 million votes

However, there are only 10 most influential politicians in the world, only 75 people with most power and influence, and only 2000 billionaires. In addition, there are 7.9 billion other inhabitants, of which 1.2 billion are youth aged 15–24 years, 16% of the global population, in 2019. The question is, is this the critical mass that can trigger a chain reaction of environmental change? To illustrate, when a nuclear chain reaction in a mass of fissile material is self-sustaining, the mass is said to be in a critical state in which there is no increase or decrease in power, temperature, or neutron population. A numerical measure of a critical mass is dependent on the

[22] https://www.theguardian.com/environment/2021/sep/28/blah-greta-thunberg-leaders-climate-crisis-co2-emissions

effective neutron multiplication factor k,[23] the average number of neutrons released per fission event that go on to cause another fission event rather than being absorbed or leaving the material. When $k = 1$, the mass is critical, and the chain reaction is self-sustaining. A supercritical mass is one which, once fission has started, will proceed at an increasing rate. The material may settle into equilibrium (i.e., become critical again) at an elevated temperature/power level or destroy itself. In the case of supercriticality, $k > 1$. Due to spontaneous fission, a supercritical mass will undergo a chain reaction. The probability that one such event will cause a chain reaction depends on how much the mass exceeds the critical mass. It needs to be noted that chain reaction is not a revolutionary process but rather an evolutionary one, a spontaneous process in which a huge amount of accumulated energy is released – in our case (the story of the environment) this is the accumulated energy of 1.2 billion young, engaged, and educated people, which are not easily manipulated (bought, bribed), and will be even less so in the future.

Wouldn't you say, then, that the situation related to environmental issues is very similar to a nuclear chain reaction? We may now be in a state of subcritical reaction when $k < 1$, but we are irresistibly approaching the moment when $k = 1$. The power, temperature, and population (approx. 1.2 billion young people) in society are growing uncontrollably, and there is not much left before a chain reaction is triggered that will lead to completely different, evolutionary social relations. Unfortunately, in our entire history, humans have still not learned how to make peaceful, evolutionary transitions. We always want a revolution, but revolutions have always shed blood and will continue to shed blood in the future as well.

13 Conclusion

This chapter provided a glimpse into the future, and in it, we emphasized that human beings are their own worst enemy but also that humans, and only humans, are part of the solution as well. Unfortunately, we are not very well aware of this, and often we consider environmental issues with our fingers crossed behind our backs. We are all aware and admit that the environmental problems are pressing and require urgent action, although not on our behalf as much as on behalf of others. But there is no "other," there is only us, all of us!

This chapter, therefore, discussed future energy and what paths we should take to preserve our way of life and at the same time to protect the environment. In the first part, the focus was on possible future scenarios for the harnessing of nuclear power (Problem 1) and, in the second part, on the problem of transport (Problem 2) and its impact on our environment. The basic starting point in presenting both problems is not in the construction and justification of scientific considerations, but rather in

[23] Effective neutron multiplication factor, k, is the average number of neutrons from one fission that cause another fission.

demonstrating how we need to build the awareness on the basis of critical thinking and concrete examples, concrete theses of pseudo-scientists and even those scientists who only tell us a partially embellished truth with the intent of manipulating arguments to deceive the public. The arguments used for the interpretation and critical judgment of certain short- and medium-term issues, claims, and concerns fall under two categories: logical, scientific arguments and arguments of the type 'pathos' and 'ethos,' which are used primarily by pseudo- and fringe scientists.

Our first conclusion was that humanity needs a constant (renewable) source of energy, which will not depend on weather conditions, such as the sun and wind. The water resources that allow us to partly achieve this have already been exploited almost to their fullest, and fossil fuels will soon run out. In a nutshell, the only alternative, at least in the medium term (by roughly 2050, when nuclear fusion may be commercially viable), is nuclear energy in the form of nuclear fission, whether we want to admit this or not. Discussions about what green energy is are philosophical in their essence; they unfold at the level of basic principles and terminology but have little to do with everyday life. It is true that nuclear energy is not absolutely clean and that it has negative impacts on the environment, but these impacts are negligible compared with any other form of energy source, even renewable ones. We need to be aware that nothing is free, that everything has to be paid one way or another!

Our second conclusion was that considering the impact of transportation on the environment, circular economy, and sustainability, an emerging trend of decarbonizing transport stands out, with the goal of making transportation systems carbon-neutral as much as possible. Achieving such an outcome requires measures that have been advocated for decades, such as low carbon fuels, vehicle and equipment efficiency, and modal shift. It remains unclear if carbon-neutral transportation is achievable in the medium term because it involves capital-intensive energy transitions. Urban transportation with a shorter lifespan of vehicles and a reliance on public transit has a better potential to become carbon-neutral. Modes such as maritime shipping have a much lower potential, mainly for technical reasons as ship engines are massive and mainly old technologies, and this level of power cannot be readily provided by technology other than the (old) internal combustion engine (McKinnon et al., 2013).

Everyone must be aware that the impacts of transportation on the environment (and the environmental footprint) are complex and multifaceted. There is a need for additional research. As better data become available, future studies should also focus on other transportation modes, those that are alternatives to automobiles and light trucks (transit), as well as air, rail, and marine, which are alternatives for long-distance travel and for freight. Today's idea of an absolute electrification of all transportation systems, however (at least until we discover truly green energy resources and appropriate electricity production methods), is ill-conceived.

Chapter 8
Conclusion

Abstract In this chapter, *Conclusion*, we finish in the same way we started: a man doesn't need an outside enemy, he's the greatest threat to himself. But he is not just a threat to himself, he's a threat also to everything and everyone around him. No living thing on earth destroys earth more than it takes for its existence, except for humans. And for that, he's willing to do anything, deceive, kill, destroy, and justify what it does with a backlash and a lie. There is a saying: "Hard things are easy; easy things are hard," called Moravec's paradox, after Hans Moravec. Moravec wrote in 1988, "it is comparatively easy to make computers exhibit adult level performance on intelligence tests or playing checkers, and difficult or impossible to give them the skills of a one-year-old when it comes to perception and mobility" (Moravec, The society of mind. Simon and Schuster, 1988, p. 15).

Similarly, Minsky emphasized that the most difficult human skills to reverse engineer are those that are unconscious. "In general, we're least aware of what our minds do best," he wrote, and added "we're more aware of simple processes that don't work well than of complex ones that work flawlessly" (Minsky, Mind children. Harvard University Press, 1986, p. 29)

The chapter concludes with the commentaries of young people to our attitudes, or especially to the attitudes of the decision-makers toward the environment. Children of the world have clearly shown what they think: "Blah, Blah, Blah." On behalf of these youngsters, Greta Thunberg criticized leaders over the climate crisis, saying:

> *They invite cherry-picked young people to meetings like this* (UNCC_COP26, 2021) *to pretend that they listen to us. But they clearly don't listen to us. Our emissions are still rising. The science doesn't lie.*
>
> *We can no longer let the people in power decide what is politically possible. We can no longer let the people in power decide what hope is. Hope is not passive. Hope is not blah, blah, blah. Hope is telling the truth. Hope is taking action. And hope always comes from the people.*

To finish in the same way we started: A man doesn't need an outside enemy, he's the greatest threat to himself. But he's not just a threat to himself, he's a threat also to everything and everyone around him. No living thing on earth destroys earth more than it takes for its existence, except for humans. And for that, he's willing to do anything, deceive, kill, destroy, and justify what it does with a backlash and a lie.

B. Aberšek, A. Flogie, *Human Awareness, Energy and Environmental Attitudes*,
https://doi.org/10.1007/978-3-031-05871-4_8

But why? There is a saying in the field of artificial intelligence: "Hard things are easy; easy things are hard," called Moravec's paradox, after Hans Moravec. The principle was articulated by Hans Moravec, Rodney Brooks, Marvin Minsky, and others in the 1980s. Moravec wrote in 1988, "it is comparatively easy to make computers exhibit adult level performance on intelligence tests or playing checkers, and difficult or impossible to give them the skills of a one-year-old when it comes to perception and mobility" (Moravec, 1988, p. 15).

Similarly, Minsky emphasized that the most difficult human skills to reverse engineer are those that are unconscious. "In general, we're least aware of what our minds do best," he wrote, and added "we're more aware of simple processes that don't work well than of complex ones that work flawlessly" (Minsky, 1986, p. 29).

One possible explanation of the paradox offered by Moravec (1988) is based on evolution. All human skills are implemented biologically using machinery designed by the process of natural selection. In the course of human evolution, natural selection has tended to preserve design improvements and optimizations. The older a skill is, the more time natural selection has had to improve the design. Abstract thought developed only very recently, and consequently, we should not expect its implementation to be particularly efficient. Activities that most people find very hard, such as playing chess or doing higher mathematics, have yielded fairly readily to computation, yet many tasks that humans find easy or even trivial resist being conquered by machines. We call the physical sciences "hard" because they deal with issues that are mostly independent of the vagaries of human nature; they offer laws that (at least in the right circumstances) yield exact answers. But physics and chemistry will never tell us how to design an effective vaccination program, prevent pollution, or solve some ethical problems of the autonomous car, in part because they do not help us comprehend human behavior. The social sciences rarely yield exact answers. But that does not make them easy. When it comes to solving real-life problems, it is the supposedly straightforward ones that seem to be tripping us up. Moravec's paradox suggests that the truly hard sciences are those that involve human behavior, as human behaviors are unpredictable and random and cannot be easily formalized, i.e., written down in the language of science, for example, in the form of some kind of mathematical model. This mathematical model would have to be – hypothetically –written down by means of a system of complex differential equations with infinite unknowns that even the most powerful computers would not be able to solve, not even using artificial intelligence. And how, then, could a human with his limited processing power, limited memory capacity, and limited abilities of critical and creative thinking be able to grasp such models, and act accordingly to them? We must take into account that humans are extremely unpredictable, which is reflected in the way we solve complex problems, such as environmental issues, or a global pandemic (Oreskes, 2020).

Development of Awareness Throughout History

Let us consider one more high-profile example, smoking. This can serve as a model example of how to develop a construal-level theory for long-term and abstract issues (such as environmental awareness). Multiple studies linked smoking to lung cancer in the late 1940s and 1950s. The medical community fiercely debated how and why smoking caused cancer in some people but not others. Before government intervention, anti-smoking efforts were led by non-profit organizations. As early as 1961, the president of the American Cancer Society and others issued a letter urging the President of the USA to create a commission that would investigate the health hazards of smoking. The 1964 landmark report *Smoking and Health: Report of the Advisory Committee to the Surgeon General of the Public Health Service* was the result. Since the 1970s, anti-smoking groups have targeted smoking in the workplace, in public places, and at home.

In the late 1970s, there was debate inside and outside the medical community about the impact of smoking on the unborn fetus and discussed the higher risk of low birth weight among other health issues for babies. Tobacco companies muddied the debate by noting the lack of clear evidence linking smoking to these problems. In this early pamphlet, the pregnant woman is encouraged to quit smoking for the health of mother and child.

By the 1990s, public health campaigns clearly stated how smoking negatively affected babies and advocated for a smoke-free environment around children. Mothers were strongly encouraged to quit smoking for the health of their children, or risk being stigmatized as a "bad mother." In the 1990s, new anti-smoking groups joined in the battle against big tobacco, using different tactics to gain support for the movement. Activists called for boycotts of tobacco companies and their affiliates, especially those known for cigarette marketing to children. Other groups used wry humor in the context of well-known cigarette advertisements, slapping public health messages onto Marlboro cowboys or Joe Camels. Some anti-smoking campaigns portrayed smoke-damaged and diseased body parts in an effort to scare smokers off cigarettes. As an example, the smoking rate in the USA has declined from 20.9% (nearly 21 of every 100 adults) in 2005 to 14.0% (14 of every 100 adults) in 2019, and the proportion of ever smokers who have quit has increased[1].

The same tactics used to cast doubt on the dangers of smoking can be used also to downplay climate change but also for COVID problems. The commentaries of young people across the world to "our" attitudes, or especially to the attitudes of the decision-makers toward the environment, have clearly shown what they think: "Blah, Blah, Blah."

On behalf of these youngsters, Greta Thunberg criticized leaders over the climate crisis, saying:

> *They invite cherry-picked young people to meetings like this to pretend that they listen to us. But they clearly don't listen to us. Our emissions are still rising. The science doesn't lie.*
>
> *We can no longer let the people in power decide what is politically possible. We can no longer let the people in power decide what hope is. Hope is not passive. Hope is not blah, blah, blah. Hope is telling the truth. Hope is taking action. And hope always comes from the people.*[2]

[1] https://www.cdc.gov/tobacco/data_statistics/fact_sheets/adult_data/cig_smoking/index.htm

[2] https://www.theguardian.com/environment/2021/sep/28/blah-greta-thunberg-leaders-climate-crisis-co2-emissions

Literature

A Technology Roadmap for Generation IV Nuclear Energy Systems. (2002). U.S. DOE Nuclear Energy Research Advisory Committee and the Generation IV International Forum.

Aberšek, B. (2012). *Didaktika tehniškega izobraževanja med teorijo in prakso* [Didactic of technical education between theory and practice]. Zavod Republike Slovenije za šolstvo.

Aberšek, B. (2013). Cogito ergo sum homomachine? *Journal of Baltic Science Education, 12*(3), 268–270.

Aberšek, B. (2016). Schola ludus. *Journal of Baltic Science Education, 15*(5), 556–558.

Aberšek, B. (2018). *Problem-based learning and proprioception*. Cambridge Scholars Publishing.

Aberšek, B. (2021). The paradox between truth and lies. *Problems of Education in the 21st Century, 79*(6), 834–836.

Aberšek, B., Borstner, B., & Bregant, J. (2014a). The virtual science teacher as a hybrid system: Cognitive science hand in hand with cybernetic pedagogy. *Journal of Baltic Science Education, 13*(1), 75–90.

Aberšek, B., Borstner, B., & Bregant, J. (2014b). *The virtual science teacher as a hybrid system*. Cambridge Scholars Publishing.

Agar, N. (2012). Why is it possible to enhance moral status and why doing so is wrong? *Journal of Medical Ethics, 2012*(39), 67–74.

Anderson, J. R. (2007). *How can the human mind occur in the physical universe*. Oxford University Press.

Anderson, L. W., & Sosniak, L. A. (1994). *Bloom's taxonomy: A forty-year retrospective*. The National Society for the Study 433 of Education.

Anderson, L. W., Krathwohl, D. R., et al. (Eds.). (2001). *A taxonomy for learning, teaching, and assessing: A revision of Bloom's taxonomy of educational objectives*. Allyn & Bacon (Pearson Education Group).

Andreiadis, S. E., Chavarot-Kerlidou, M., Fontecave, M., & Artero, V. (2011). Artificial photosynthesis: From molecular catalysts for light-driven watersplitting to photoelectrochemical cells. *Photochemistry and Photobiology, 2011*(87), 946–964.

Banister, D. (2008). The sustainable mobility paradigm. *Transport Policy, 15*(2), 73–80.

Banister, D., & Button, K. (Eds.). (1993). *Transport, the environment, and sustainable development*. Spon Press.

Bermudez, J. L. (2010). *Cognitive Science*. Cambridge University Press.

Black, W. R. (2010). *Sustainable transportation: Problems and solutions*. The Guilford Press.

Beck, U. (1992). Risk Society: Towards a New Modernity. Sage.

B. Aberšek, A. Flogie, *Human Awareness, Energy and Environmental Attitudes*, https://doi.org/10.1007/978-3-031-05871-4

Bloom, B. S., Engelhart, M. D., Furst, E. J., Hill, W. H., & Krathwohl, D. R. (1956). *Taxonomy of educational objectives: The classification of educational goals. Handbook I: Cognitive domain*. David McKay Company.

Bohm, D. (2004). *On dialog*. Routledge Clasics.

Bonz, B., & Ott, B. (2003). *Algemeine Technikdidaktik – Theoriensatze und Praxisbezüge*. Schneider Verlag Hohengehren GmbH.

Bostrom, N. (2014). *Superintelligence: Paths, dangers, strategies*. Oxford University Press.

Bradford, T. (2018). *The energy system: Technology, economics, markets, and policy*. The MIT Press.

Brasseur, G. P., et al. (2016). Impact of aviation on climate: FAA's aviation climate research initiative (ACCRI) phase II. *Bulletin of the American Meteorological Society, 97*(4), 561–583. https://doi.org/10.1175/BAMS-D-13-00089.1

Bryant, D. A., & Frigaard, N. U. (2006). Prokaryotic photosynthesis and phototrophy illuminated. *Trends in Microbiology, 14*(11), 488–496. https://doi.org/10.1016/j.tim.2006.09.001

Bryson, Y. M. (2018). *Strategic planning for public and nonprofit organizations: A guide to strengthening and sustaining organizational achievement*. Willy.

Calzaferri, G. (2010). Artificial photosynthesis. *Topics in Catalysis, 53*(3), 130–140.

Camus, A. (1999). *The myth of sisyphus*. Penguin Books.

Chalmers, D. (1996). *The conscious mind: In search of a fundamental theory*. Oxford University Press.

Chapin, F. S., Matson, P. A., & Mooney, H. A. (2002). *Principles of terrestrial ecosystem ecology* (pp. 97–104). Springer. ISBN:978-0-387-95443-1.

Chapman, J. D. (1989). *Geography and energy: Commercial energy systems and national policies*. Longman Scientific & Technical.

Cold fusion: A case study for scientific behavior. https://undsci.berkeley.edu/article/cold_fusion_01

Dawkins, R. (2017). *The selfish gene*. Macat Library.

Dennett, D. (2012). *From bacteria to Bach and back: The evolution of minds*. Penguin.

Dignum, V. (2019). *Responsible Artificial Intelligence*. Springer.

Dogutan, D. K., & Nocera, D. N. (2019). *Artificial photosynthesis at efficiencies greatly exceeding that of natural photosynthesis*. American Chemical Society.

Donnison, C., et al. (2020). Bioenergy with Carbon Capture and Storage (BECCS): Finding the win–wins for energy, negative emissions and ecosystem services—size matters. *GCB Bioenergy, 12*, 586–604.

Drummond, C., & Fischhoff, B. (2017). Development and validation of the scientific reasoning scale. *Journal of Behavioral Decision Making, 30*(1), 26–38.

European Commission. (2004). *European fusion research area*. Office for Official Publications of the European Communities.

European Commission. (2015a). *Paris climate conference: Historic opportunity to avoid dangerous climate change*. https://ec.europa.eu/commission/presscorner/detail/en/IP_15_6160

European Commission. (2015b). *Historic climate deal in Paris: EU leads global efforts*. https://ec.europa.eu/commission/presscorner/detail/en/IP_15_6308

European Commission. (2015c). Historic climate deal in Paris: EU leads global efforts. https://ec.europa.eu/commission/presscorner/detail/en/IP_15_6308

European Commission. (2018a). *2050 energy strategy*. Pridobljeno 15.5.2018s. https://ec.europa.eu/energy/en/topics/energy-strategy-and-energy-union/2050-energy-strategy

European Commission. (2018b). *2050 low-carbon economy*. Pridobljeno 16.5.2018s. https://ec.europa.eu/clima/policies/strategies/2050_sl

European Commission. (2018c). *COP21 UN climate change conference, Paris*. Pridobljeno 13.4.2018s. https://ec.europa.eu/commission/priorities/energy-union-and-climate/climate-action-decarbonising-economy/cop21-un-climate-change-conference-paris_en

European Commission. (2018d). *Effort sharing 2021–2030: Targets and flexibilities*. Pridobljeno 12.8.2018s. https://ec.europa.eu/clima/policies/effort/proposal_en

Fodor, J. (1975). *The language of thought*. Harvard University Press.

Fodor, J. A. (1983). *Modularity of mind: An essay on faculty psychology*. The MIT Press.
Fodor, J. (1987). *Psychosemantics*. The MIT Press.
Fujishima, A., Rao, T. N., & Tryk, D. A. (2000). Titanium dioxide photocatalysis. *Journal of Photochemistry and Photobiology C: Photochemistry Reviews, 1*(1), 1–21. https://doi.org/10.1016/S1389-5567(00)00002-2
Gilbert, R., & Perl, A. (2008). *Transport revolutions. Moving people and freight without oil*. Earthscan.
Goldman, T., & Gorham, R. (2006). Sustainable urban transport: Four innovative directions. *Technology in Society, 28*, 261–273.
Goleman, D. (1995). *Emotional intelligence*. A Bantam Book.
Goleman, D. (2007). *Social intelligence*. Arrow.
Govindjee, R. (2014). *What is photosynthesis?* Biology at Illinois.
Green, B. (2005). *The fabric of the cosmos: Space, time, and the texture of reality*. Alfred A. Knopf.
Grimes, D. R. (2021). *COVID has created a perfect storm for fringe science*. Scientific American.
Hempel, C. G. (1965). *Aspects of scientific explanation and other essays in the philosophy of science*. The Free Press.
Hempel, C. G., & Oppenheim, P. (1948). Studies in the logic of explanation. *Philosophy of Science, 15*, 135–175.
Hickman, R., Hall, P., & Banister, D. (2013). Planning more for sustainable mobility. *Journal of Transport Geography, 33*, 210–219.
Hirsch, H., & Weish, P. (2007). Environmental pollution caused by the 'Normal operation' nuclear fuel cycle. In *LEBENSMINISTERIUM 2007* (pp. 43–52).
Hirsch, H., Becker, O., Schneider, M., & Froggatt, A. (2005). *Nuclear reactor hazards: Ongoing dangers of operating nuclear technology in the 21st century*. Report prepared for Greenpeace International. www.greenpeace.org/international/press/reports/nuclearreactorhazards
Hong, J., Dong, K., Chan, H. N., & Park, B. (2014). Nanobiocatalytic assemblies for artificial photosynthesisel. *Current Opinion in Biotechnology, 28*, 1–9.
Huizenga, J. (1993). *Cold fusion: The scientific fiasco of the century*. University of Rochester Press.
IAEA PRIS. (2007). *Annual report 2007*. International Atomic Energy Agency. https://www.iaea.org/sites/default/files/publications/reports/2007/anrep2007_full.pdf
IEA. (2019). *The future of rail: Opportunities for energy and the environment*. IEA. https://doi.org/10.1787/9789264312821-en
IMO (2015). Third IMO Greenhouse Gas Study 2014: Executive Summary and Final Report. international Maritime Organization. https://greenvoyage2050.imo.org/wp-content/uploads/2021/01/third-imo-ghg-study-2014-executivesummary-and-final-report.pdf
ITEA. (2003). *Advancing excellence in technological literacy: Students assessment, professional development and program standard*. International Technology Education Association.
ITEA. (2007). *Standard for technological literacy, content for the study of technology*. International Technology Education Association.
Jiao, F., & Frei, H. (2009). Nanostructured cobalt oxide clusters in mesoporous silica as efficient oxygen-evolving catalysts. *Angewandte Chemie, 48*(10), 1841–1844.
JCAP (2020). 2020 Solar fuel science meeting. The joint center for artificial photosynthesis (JCAP) https://solarfuelshub.org/2020-solar-fuels-science-meeting
Jones, S. E., Palmer, E. P., Czirr, J. B., Decker, D. L., Jensen, G. L., Thorne, J. M., Taylor, S. F., & Rafelski, J. (1989). Observation of cold nuclear fusion in condensed matter. *Nature, 388*, 737–740.
Kahneman, D. (2011). *Thinking, fast and slow*. Farrar, Straus and Giroux.
Khoulya, M. E., Mohsnawy, E., & Fukuzumic, S. (2017). Solar energy conversion: From natural to artificial photosynthesis. *Journal of Photochemistry and Photobiology C: Photochemistry Reviews, 31*, 36–83.
Kim, J. (1996). *Philosophy of mind*. Westview Press.
Kim, J. (2003). The American origins of philosophical naturalism. *Journal of Philosophical Research, 28*(September), 83–98.

Kordigel Aberšek, M., & Aberšek, B. (2020). *Society 5.0 and literacy 4.0 for 21st century*. Nova Science Publishers.
Krikorian, Y. H. (1944). *Naturalism and the human spirit*. Columbia University Press.
Kutscher, C. F., Milford, J. B., & Kreith, F. (2018). *Principles of sustainable energy systems* (3rd ed.). CRC Press.
Lamar Keene, M., & Spraggett, A. (1976). *The psychic mafia*. St. Martin's Press.
Lee, D. S., et al. (2021). The contribution of global aviation to anthropogenic climate forcing for 2000 to 2018. *Atmospheric Environment, 244*, 117834.
Lehninger, A. L., Nelson, D. L., & Cox, M. M. (2008). *Principles of biochemistry* (5th ed.). Example Product Manufacturer.
Leiserowitz, A., Carman, J., Buttermore, N., Wang, X., Rosenthal, S., Marlon, J., & Mulcahy, K. (2021). *International public opinion on climate change*. Yale Program on Climate Change Communication and Facebook Data for Good.
Lenk, H., & Ropohl, G. (1989). *Systemtheorie als Wissenschaftsprogramm*. Athenaeum Vlg.
Lowe, M. D. (1990). Alternatives to the automobile: Transport for livable cities. *Ekistics, 344*(345), 269–282.
Lowy, J. (2016). *UN agreement reached on aircraft climate-change emissions*. Associated Press.
Markič, O. (2010). *Kognitivna znanost*. Aristej.
Maxwell, K., & Johnson, G. N. (2000). Chlorophyll fluorescence – A practical guide. *Journal of Experimental Botany, 51*(345), 659–668. https://doi.org/10.1093/jexbot/51.345.659
McKinnon, A. (2018a). *Balancing efficiency and resilience in multimodal supply chains* (International Transport Forum Discussion Papers). OECD Publishing.
McKinnon, A. (2018b). *Decarbonizing logistics: Distributing goods in a low carbon world*. Kogan Page.
McKinnon, A., Browne, M., & Whiteing, A. (Eds.). (2013). *Green logistics: Improving the environmental sustainability of logistics* (2nd ed.). Kogan Page.
Minsky, M. (1986). *The society of mind*. Simon and Schuster.
Miyamoto, K. (1994). In Y. Murooka & T. Imanaka (Eds.), *Recombinant microbes for industrial and agricultural applications* (pp. 771–785). Marcel Dekker.
Miyamoto, K. (2013). Chapter 1: Biological energy production. In *Renewable biological systems for alternative sustainable energy production* (FAO Agricultural Services Bulletin – 128). Food and Agriculture Organization of the United Nations.
Moravec, H. (1988). *Mind children*. Harvard University Press.
Müller, K. H. (2008). *The new science of cybernetics. The evolution of living research designs, Vol. I: Methodology*. Edition Echoraum.
Müller, K. H. (2011). *The new science of cybernetics. The evolution of living research designs, Vol. II: Theory*. Edition Echoraum.
NAEP. (2013). *National assessment of educational progress*. Retrieved from: https://catalog.data.gov/dataset/2013-national-assessment-of-educational-progre
NAEP. (2014). *Technology and engineering literacy framework for the 2014 national assessment of educational progress*. Retrieved from: https://nagb.gov/content/nagb/assets/documents/publications/frameworks/technology/2018-technology-framework.pdf
NAEP. (2018). *Technology and engineering literacy assessment*. Retrieved from: https://www.nationsreportcard.gov/tel/
NAGB. (2018). *Technology and engineering literacy framework for the 2018 national assessment of educational progress*. National Assessment Governing Board. Retrieverd form: www.nagb.gov
Nawaz, A., Kuila, A., Rani, A., Mishra, N.S., Sim, L.C., Leong, K.H. and Saravanan, P. (2019). Industrial application of light-driven nanomaterial. Pages 151–179, Editor(s): Sabu, T., Grohens, Y. and Pottathara, Y.B. In Micro and Nano technologies, industrial applications of nanomaterials. Elsevier
NERAC (2002). A Technology Roadmap for Generation IV Nuclear Energy Systems. U.S. DOE Nuclear Energy Research Advisory Committee and the Generation IV International Forum.

Nersesian, R. L. (2016). *Energy economics: Markets, history and policy*. Routledge.
Newell, A., & Simon, H. (1972). *Human problem solving*. Prentice Hall.
Noussan, M., Hafner, M., & Tagliapietra, S. (2020). *The future of transport between digitalization and decarbonization: Trends, strategies and effects on energy consumption*. Springer Briefs in Energy.
Obersteiner, M. (2001). Managing climate risk. *Science, 294*(5543), 786–787.
OECD. (2005). *The definition and selection of key competencies*. OECD Publishing.
OEKO. (2007). *Öko -Institute 2007 annual report*. Öko-Institut e.V., Institute for applied Ecology.
Online Etymology Dictionary. (2020). *Photosynthesis*. http://654.pl/lp/wiki/view.php?sq=streaming&lang=en&q=Photosynthesis
Oppenheim, P., & Putnam, H. (1958). Unity of science as a working hypothesis. *Minnesota Studies in the Philosophy of Science, 2*, 3–36.
Oreskes, N. (2020, December). History matters to science. *Scientific American*.
Pang, R., Teramura, K., Morishita, M., Asakura, H., Hosokawa, S., & Tanaka, T. (2020). Enhanced CO evolution for photocatalytic conversion of CO_2 by H_2O over Ca modified Ga_2O_3. *Communications Chemistry, 3*, Article Number: 137.
Parisi, J., & Ball, J. (2019). *The future of fussion energy*. World Scientific.
PIRLS. (2016). *International results in reading*. Retrieved from: http://pirls2016.org/download-center/
PISA. (2015). *Pisa 2015: Draft reading literacy framework*. OECD Publications Service.
PISA. (2018). *OECD program for international student assessment*. OECD Publications Service.
Quader, M. A., & Ahmed, S. (2017). Chapter 4: Bioenergy with carbon capture and storage (BECCS): Future prospects of carbon-negative technologies. In M. G. Rasul, A. K. Azad, & S. Sharma (Eds.), *Clean energy for sustainable development comparisons and contrasts of new approaches*. Academic.
Rahwan, I. (2018). Society-in-the-loop: Programming the algorithmic social contract. *Ethics and Information Technology, 20*, 5–14.
Rasul, M. G., Azad, A. K., & Sharma, S. C. (2017). *Clean energy for sustainable development comparisons and contrasts of new approaches*. Academic.
Raven, P. H., Evert, R. F., & Eichhorn, S. E. (2005). *Biology of plants* (7th ed., pp. 124–127). W. H. Freeman and Company. ISBN:978-0-7167-1007-3.
Reece, S. Y., Hamel, J. A., Sung, K., Jarvi, T. D., Esswein, A. J., Pijpers, J. J. H., & Nocera, D. G. (2011). Wireless solar water splitting using silicon-based semiconductors and earth-abundant catalysts. *Science, 334*(6056), 645–648.
Schiller, P. L., & Kenworthy, J. R. (2018). *An introduction to sustainable transportation: Policy, planning and implementation*. Routledge.
Schulenberg, T., Behnke, L., Hofmeister, J., & Loewenberg, M. (2004). *What is generation IV?* Forschungszentrum Karlsruhe.
Searle, J. R. (1980/1990). Minds, brains and programs. In M. A. Boden (Ed.), *The philosophy of artificial intelligence* (pp. 67–88). Oxford University Press.
Searle, J. R. (1983). *Intentionality: An essay in the philosophy of mind*. Cambridge University Press.
Semrud-Clikeman, M. (2007). Social competence in children. In *Social competence in children*. Springer. https://doi.org/10.1007/978-0-387-71366-3_1
Swedish Consortium for Artificial Photosynthesis. (2020). Uppsala University. Archived from the original on 20 January 2012. Retrieved 1.10.2020.
Tolley, R. (Ed.). (2003). *Sustainable transport: Planning for walking and cycling in urban environments*. CRC Press.
Trope, Y., & Liberman, N. (2010). Construal-level theory of psychological distance. *Psychological Review, 117*(2), 440–463. https://doi.org/10.1037/a0018963
UNCC. (2018). *Kyoto protocol introduction*. Pridobljeno 26.2.2018s. https://unfccc.int/process/the-kyoto-protocol
UN-HABITAT (2009). Global report on human settlements 2009: Planning sustainable cities - Abridged Edition.

UNCC_COP26. (2021). *United Nations Climate Change Conference COP26 (2021)*. https://www.un.org/en/climatechange/cop26

UNESCO. (2002). *Technical and vocational education and training for the twenty-first century: UNESCO and ILO recommendations*. UNESCO, ILO.

UNESCO. (2008). *Education for all global monitoring report 2009. Overcoming inequality: Why governance matters*. UNESCO.

UNFCCC. (1992). *United Nations framework convention on climate change*. United Nations, General Assembly.

Van Noorden, R. (2012). 'Artificial leaf' faces economic hurdle: Nature news & comment. *Nature*.

Venturi, M., Balzani. V., & Gandolfi, M. T. (2005). *Fuels from solar energy. A dream of Giacomo Ciamician, the father of photochemistry*. In American Solar Energy Society in the Proceedings of the 2005 Solar World Congress Orlando, FL, USA.

Wang, H., et al. (2020a). Acclimation of leaf respiration consistent with optimal photosynthetic capacity. *Global Change Biology, 26*(4), 2573–2583.

Wang, Q., Warnan, J., Rodríguez-Jiménez, S., et al. (2020b). Molecularly engineered photocatalyst sheet for scalable solar formate production from carbon dioxide and water. *Nature Energy, 5*, 703–710. https://doi.org/10.1038/s41560-020-0678-6

Wealer, B., Bauer, S., Landry, N., Seiß, H., & von Hirschhausen, C. R. (2018). *Nuclear power reactors worldwide: Technology developments, diffusion patterns, and country-by-country analysis of implementation (1951–2017)* (DIW Data Documentation, No. 93). Deutsches Institut für Wirtschaftsforschung (DIW).

Weber, M. (2021, April). Can natural gas be part of a low-carbon future? *Scientific American*.

Wenisch, A., & Prauhart, N. (2006). Chapter 5: Die Renaissance der Kernenergie – Wunschtraum der Atomlobby oder Realität? aus *50 Jahre Atomenergie sind genug. Zum Status der Atomkraft im 21. Jahrhundert* (pp. 45–53). Ökologie-Institut (Austrian Institute of Ecology).

Wenisch, A., Kromp, R., & Reinberger, D. (2007). *Science or fiction: Is the future for nuclear?* Austrian Institute of Ecology.

World Energy Council. (2007). *Transport technologies and policy scenarios*. World Energy Council.

Yarris, L. (2009). *Turning sunlight into liquid fuels: Berkeley lab researchers create a nano-sized photocatalyst for artificial photosynthesis*. Berkeley Lab News Center. Lawrence Berkeley National Laboratory.

Index

B. Aberšek, A. Flogie, *Human Awareness, Energy and Environmental Attitudes*,
https://doi.org/10.1007/978-3-031-05871-4